START MAKING!

A Guide to Engaging Young People in Maker Activities

BY DANIELLE MARTIN AND ALISHA PANJWANI

Natalie Rusk, Editor

SAN FRANCISCO, CA

Start Making!
A Guide to Engaging Young People in Maker Activities

By Danielle Martin and Alisha Panjwani
Natalie Rusk, Editor

Published by Maker Media, Inc.,
1160 Battery Street East, Suite 125,
San Francisco, California 94111.

Maker Media books may be purchased for educational, business, or sales promotional use. Online editions are also available for most titles (safaribooksonline.com/). For more information, contact our corporate/institutional sales department: 800-998-9938 or corporate@oreilly.com.

Publisher: Roger Stewart
Editor: Roger Stewart
Copy Editor and Proofreader: Rebecca Rider, Happenstance Type-O-Rama
Interior and Cover Designer: Maureen Forys, Happenstance Type-O-Rama
Indexer: Valerie Perry, Happenstance Type-O-Rama

April 2016: First Edition
Revision History for the First Edition
2016-04-05: First Release

See oreilly.com/catalog/errata.csp?isbn=9781457187919 for release details.

Safari® Books Online

Safari Books Online is an on-demand digital library that delivers expert content in both book and video form from the world's leading authors in technology and business.

Technology professionals, software developers, web designers, and business and creative professionals use Safari Books Online as their primary resource for research, problem solving, learning, and certification training.

Safari Books Online offers a range of plans and pricing for enterprise, government, education, and individuals. Members have access to thousands of books, training videos, and prepublication manuscripts in one fully searchable database from publishers like O'Reilly Media, Prentice Hall Professional, Addison-Wesley Professional, Microsoft Press, Sams, Que, Peachpit Press, Focal Press, Cisco Press, John Wiley & Sons, Syngress, Morgan Kaufmann, IBM Redbooks, Packt, Adobe Press, FT Press, Apress, Manning, New Riders, McGraw-Hill, Jones & Bartlett, Course Technology, and hundreds more. For more information about Safari Books Online, please visit us online.

How to Contact Us

Please address comments and questions concerning this book to the publisher:

Make:
1160 Battery Street East, Suite 125
San Francisco, CA 94111
877-306-6253 (in the United States or Canada)
707-639-1355 (international or local)

Make: unites, inspires, informs, and entertains a growing community of resourceful people who undertake amazing projects in their backyards, basements, and garages. Make: celebrates your right to tweak, hack, and bend any technology to your will. The Make: audience continues to be a growing culture and community that believes in bettering ourselves, our environment, our educational system—our entire world. This is much more than an audience, it's a worldwide movement that Make is leading we call it the Maker Movement.

For more information about Make:, visit us online:

• Make: magazine makezine.com/magazine
• Maker Faire makerfaire.com
• Makezine.com makezine.com
• Maker Shed makershed.com

To comment or ask technical questions about this book, send email to bookquestions@oreilly.com.

Contents

PART I Get Ready to Start Making!

PART II Start Making! Sessions

PART III Keep Making

Foreword

A few years ago a small team of us at Intel developed an outreach program drawing on the skills and passions of our resident makers, which we called "Start Making!" Our aims were to complement Intel's ongoing efforts to inspire students in science, technology, engineering, and math (STEM) fields and to attract a more diverse population of youth to consider educational and career pathways in technology.

Start Making! workshop (HENNEPIN COUNTY LIBRARY BEST BUY TEEN TECH CENTER, MINNEAPOLIS, MN)

WHAT WILL *YOU* MAKE?

We believe that great numbers of young people out there—some of whom, for one reason or another, do not necessarily self-identify as strong in engineering or design—can and will make major creative contributions toward building our (necessarily technological) future. Furthermore, we believe we must reach those young makers through nontraditional channels. We can open the doors to creative careers in high tech and help minimize the barriers to entry by

▶ Eliminating the intimidation factors that some students may associate with STEM subjects

▶ Highlighting the "hooks" that will appeal not only to the mechanically- and mathematically-inclined novice makers, but also to those who are naturally gifted in expression through textile arts, spatial arts, performance arts, music, and so on

▶ Offering tools that enable immediate success and providing environments that support inclusivity, open learning, and creative exploration

The Maker movement—a recent wave of tech-inspired, do-it-yourself (DIY) innovation—is sweeping the globe. Participants in this movement, known as *makers*, take advantage of cheap, powerful, easy-to-use tools, as well as easier access to knowledge, capital, and markets, to create new physical objects. This revolutionary change in how hardware is innovated and manufactured has great potential to change the future of computing, particularly for young people from backgrounds

Intel's Jay Melican with Clubhouse youth at the Bay Area Maker Faire

traditionally underrepresented in STEM fields: females, racial and ethnic minority groups, and people with disabilities.

By empowering girls and young people from other underrepresented groups to just "Start Making!" we can open the doors to technological innovation (and to the recognition of potential career opportunities in high tech) for a large and extraordinarily talented crowd of young makers who may otherwise be locked out by traditional STEM education programs.

Four years after we started down this path at Intel, we are thrilled to see that Start Making! has grown, matured, and evolved under the expert guidance of The Clubhouse Network team, in collaboration with the Lifelong Kindergarten group at the MIT Media Lab. The original spirit behind the effort to educate a generation of makers has been amplified as more and more creative, talented, local facilitators have customized the program to engage youth from their communities. Through a knowledge-sharing network of almost 100 Clubhouses, The Clubhouse Network is enabling thousands of young people to practice "making" in their daily lives. This book aims to

Start Making! Clubhouse Coordinator workshop, Denver, CO

expand that reach and broaden the community of young makers even further by sharing these ideas and approaches.

The Start Making! activities described in this book are undeniably fun; they also present young learners and makers with some of the foundational building blocks for understanding electronics and computing, such as how a circuit works or what it means for a material to be conductive. More importantly, these activities set the stage for beginner makers to do something creative and original with that knowledge. The activities come from the maker community and capture the wonderful spirit of open collaboration, self-directed learning, and fun.

Intel volunteers learning soft circuits side by side with Clubhouse youth (BOYS & GIRLS CLUBS OF EAST VALLEY, CHANDLER, AZ)

As the name suggests, Start Making! is just the beginning. We now see easy-to-use coding environments and physical computing platforms made widely available, empowering makers of all skill levels to design interactive objects that sense and control the physical world around them.

It is our hope that Start Making! will inspire young learners to keep making—to pick up new tools because they have gained confidence through these activities and know that technology can be a powerful means of creative expression, and that technological devices are not just things you can buy, but things you can learn to build yourself to create a better world.

—Carlos Contreras (Public Affairs Director, Intel)

—Anne McGrath (Program Manager, Intel Foundation)

—Jay Melican (Maker Czar, Intel)

Acknowledgments

This guide builds on the curriculum developed by Karen Tanenbaum during her work at Intel. The guide also builds on activities and ideas from Leah Buechley, Mike Petrich, Jie Qi, Jay Silver, and Karen Wilkinson.

Other contributors include Mitchel Resnick, Jackie Gonzalez, and Clubhouse staff and mentors around the world. A special thank you to all of the Clubhouse youth throughout the Network who tested these activities and helped to make them even more engaging, meaningful, and fun.

Thanks also to Gail Breslow, Director of the Clubhouse Network, for supporting and guiding the development of this book and the Start Making! initiative.

Roger Stewart at Maker Media expertly guided the book from draft to completion, and Michelle Hlubinka of Maker Media provided valuable feedback along the way.

We are grateful for support from Intel, the LEGO Foundation, and the Lemann Foundation.

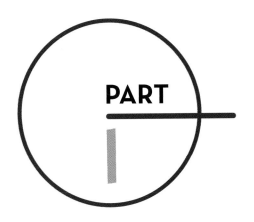

PART

I

Get Ready to Start Making!

Welcome to
START MAKING!

Around the world, children and teens are becoming engaged in making. They are designing light-up cards for family and friends, building machines that draw, programming musical instruments, and creating their own toys using recycled parts. In the process of designing projects, they are learning new ways to solve problems, communicate ideas, and collaborate with others.

Exploring electronics (FARO DE ORIENTE CLUBHOUSE, MEXICO CITY, MEXICO)

Making design journals (FLAGSHIP CLUBHOUSE, MUSEUM OF SCIENCE, BOSTON, MA)

This guide provides ideas and activities that you can use to help young people start making. In this guide, we share Start Making!—a program that has engaged hundreds of youth in the process of designing their own projects.

This Start Making! guide offers a series of creative do-it-yourself (DIY) projects that introduce young people to the basics of circuitry, coding, crafting, and engineering. Starter project activities lead into Open Make sessions during which young people work on personalized projects, both on their own and in small groups. Through the process of designing and making projects, young people build confidence, camaraderie, and curiosity about science, technology, art, engineering, and math concepts.

Start Making! consists of a series of activity sessions that you can adapt to your situation. You can offer your own version of Start Making! activities in your home, at the library, at an after-school club, at the local community center, or anywhere else young people can gather to work on projects together. You can dip into the activities once a week, run them as a week-long summer activity, or go through them in any way that works for you and your group.

We developed the Start Making! program within The Clubhouse Network, a global network of community-based centers where youth create projects based on their interests using a variety of tools and technologies. Facilitators provide support and model the process of making projects. Throughout this guide, we share examples of Start Making! projects and experiences from Clubhouse youth (ages 10 to 15) and facilitators around the world. You can learn more about Clubhouses on pages 185-187.

We encourage you to take these ideas and make them your own. We hope this guide will help you create more opportunities for young people to start making projects together. By offering your own Start Making! program, you can inspire young people in your community to develop creative ideas, learn new skills, and share their creations.

WHAT IS MAKING?

We define *making* as the process of creating projects based on your ideas and interests. We encourage a playful and curious approach to the process.

The making activities we share in this book bridge concepts and techniques from art, crafts, music, and design with science, technology, engineering, and math—an integration of ideas referred to by some educators as STEAM (science, technology, engineering, art, and math), suggesting the power of this integration for motivating learning. The projects mix familiar materials (such as paper, fabric, and recycled materials) with new conductive and programmable materials (such as LEDs, conductive thread, and microcontrollers). We have found that more young people become interested in science and technology concepts when the concepts are applied to making projects that integrate art, music, and design.

Testing a circuit (TECNOCENTRO SOMOS PACÍFICO CLUBHOUSE, CALI, COLOMBIA)

WHO IS A MAKER?

We believe everyone has the potential to be a maker: to be inspired to imagine, create, and share personally meaningful projects. The Start Making! program is designed to help young people begin to identify themselves as makers. We recognize that young people are much more likely to see themselves as makers when they feel part of a creative community.

Making a light-up circuit with support from a facilitator (FLAGSHIP CLUBHOUSE)

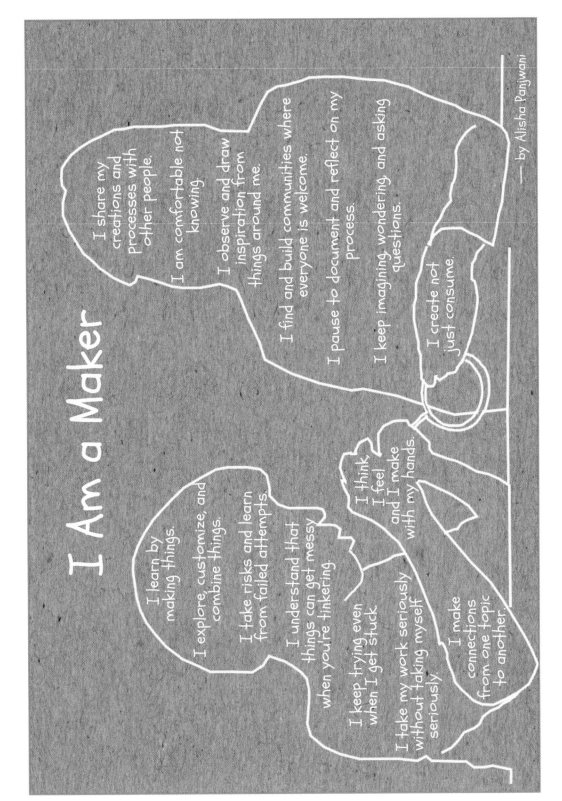

I Am a Maker

I share my creations and processes with other people.

I am comfortable not knowing.

I observe and draw inspiration from things around me.

I find and build communities where everyone is welcome.

I pause to document and reflect on my process.

I keep imagining, wondering, and asking questions.

I create not just consume.

I think I feel and I make with my hands.

I learn by making things.

I explore, customize, and combine things.

I take risks and learn from failed attempts.

I understand that things can get messy when you're tinkering.

I keep trying even when I get stuck.

I take my work seriously without taking myself seriously.

I make connections from one topic to another.

— by Alisha Panjwani

GUIDING PRINCIPLES

Start Making! is based on four guiding principles. These principles grew out of research by the Lifelong Kindergarten group at the MIT Media Lab, and they form the core of the Clubhouse learning model.

Principle 1: Support learning through design experiences. The Start Making! program is based on the idea that people learn best when they are engaged in creating personally meaningful products. As young people work on projects, they can be seen as engaging in a design process, which we call a "creative learning spiral" (see the following image). In this process, they *imagine* what they want to do, *create* a project based on their ideas, *play* with alternatives, *share* their ideas and creations with others, and *reflect* on their experiences—all of which lead them to *imagine* new ideas and new projects. As youth engage in these experiences, they learn valuable technical skills while also learning about the process of design and invention.

Principle 2: Help youth build their interests. When young people care about what they are working on, they are willing to work longer and harder, and they learn more in the process. In Start Making! facilitators help young people gain experience with self-directed learning, providing support for youth to recognize, trust, develop,

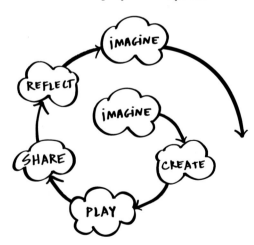

The creative learning spiral (CREDIT: MITCHEL RESNICK AND NATALIE RUSK, LIFELONG KINDERGARTEN GROUP)

and deepen their own interests and talents. Many youth begin by mimicking a sample project, then work on variations on the theme, and soon develop their own personal path, stemming from their personal interests.

Principle 3: Develop a sense of community. Start Making! is also designed to develop a learning community in which youth share ideas and work together on projects. Facilitators play an important role not just in supporting youth, but also by modeling the process of making and learning themselves.

Principle 4: Foster an environment of respect and trust. In Start Making! programs, young people are treated with trust and respect—and are expected to treat others the same way. Start Making! facilitators strive to create an environment in which participants feel safe to experiment, explore, and innovate.

GOALS OF START MAKING!

Start Making! is designed to encourage young people to develop their own ideas, to experiment, and to innovate. Teaching young people the activities themselves is not the primary goal. Rather, the goal is to enable young people to develop their own projects and to foster motivation and confidence in their ability to learn.

By offering Start Making! you can help young people develop creative competencies. Here are the five key competencies that we identified as outcomes for young people who participate in Start Making!

Identify as a creator or maker. Young people develop positive attitudes toward creating hands-on projects.

Develop confidence in creative expression. Young people feel capable of bringing their ideas to life by designing, experimenting, iterating, and persisting through failures.

Acquire technical tool literacy. Young people become familiar with a variety of tools and technologies that they can use to make projects.

Become aware of STEAM. Young people become aware of ideas and concepts that bridge science, technology, engineering, art, and math and demonstrate curiosity to learn more.

Learn collaboration and networking skills. Young people actively engage in collaborating and helping others.

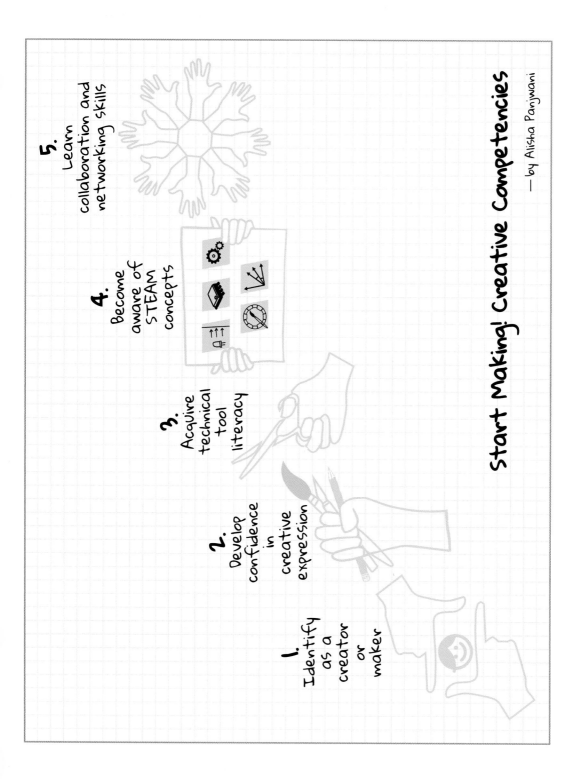

Start Making! Creative Competencies

— by Alisha Panjwani

1. Identify as a creator or maker

2. Develop confidence in creative expression

3. Acquire technical tool literacy

4. Become aware of STEAM concepts

5. Learn collaboration and networking skills

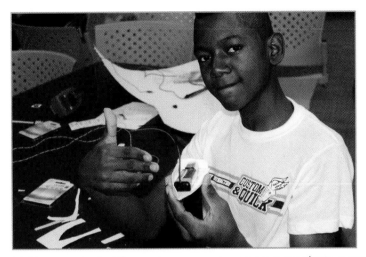

Making a project with a motor (TECNOCENTRO SOMOS PACÍFICO CLUBHOUSE)

An evaluation of the Start Making! program in Clubhouses showed that young people gained confidence and experience in each of these five areas. (To learn more, see the evaluation report by Julie Remold of SRI International at bit.ly/start-making-evaluation-report).

HOW DO I DESIGN MY OWN START MAKING! PROGRAM?

The Start Making! program consists of a series of sessions that you can adapt to your local context. For example, you can offer it as a full-day camp over the course of a week, or as an after-school program that meets for two to three hours each session.

The first six sessions introduce new tools and techniques through starter projects. Each of these sessions includes Open Make time, during which young people apply what they've learned to create personalized projects. These core sessions lead into a final Open Make session in which the makers spend time preparing a final project. The program culminates in a showcase where they share their projects with family, friends, and other community members.

The projects and activities in the first six sessions help makers understand simple concepts and then dive deeper into more complex ideas. Over the course of multiple sessions, young makers learn to develop new ideas, persist through setbacks, collaborate with others, and make meaningful projects.

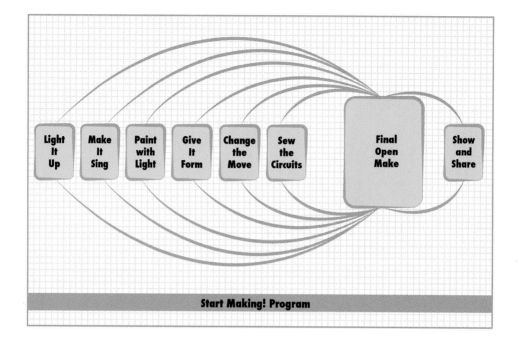

Start Making! Program

Here's a brief description of each session:

Session 1. Light It Up: Paper Circuits Introduces basic circuits through making light-up cards and other creations using LEDs and copper tape applied to paper.

Session 2. Make It Sing: Programmable Musical Creations Introduces physical computing and programming through the process of making interactive musical instruments using Scratch and the MaKey MaKey invention kit.

Session 3. Paint with Light: Illuminated Wands and Photography Introduces light painting using LEDs and delayed-exposure photography.

Session 4. Give It Form: 3D Forms Introduces three-dimensional spatial design through the process of creating a digital sculpted character.

Session 5. Change the Move: Art-Making Bots Introduces reverse engineering and remixing by deconstructing a toothbrush or other simple device to build art-making robots.

Session 6. Sew the Circuits: E-Textiles Introduces e-textiles by designing a sewable project, such as a bracelet, pin, or flag, using a preprogrammed LilyTiny board.

Session 7. Final Open Make: Personalized Projects Encourages young people to develop projects, individually or in groups, based on their personal interests, applying techniques explored in earlier sessions.

Session 8. Show and Share: Community Showcase Provides a showcase in which young people can share their projects and process, celebrating their accomplishments with family and friends.

You can offer all of the six core sessions or choose the ones you think will work best for your group. After the core sessions, provide Open Make time for preparing final projects. Then, collaborate with young people to organize a community event for project sharing.

YOUR ROLE AS A FACILITATOR

As a facilitator, your role is to create a welcoming environment in which people feel encouraged to imagine new ideas and bring them to life through creative hands-on explorations. Rather than directly instructing, your focus as a facilitator is on providing a safe and inviting space in which makers can experiment and learn from their own explorations.

We suggest finding others who can join you to help facilitate your Start Making! program as volunteers or staff. We've found the most important qualities for facilitators are an interest in working with young people and a passion for making and learning themselves.

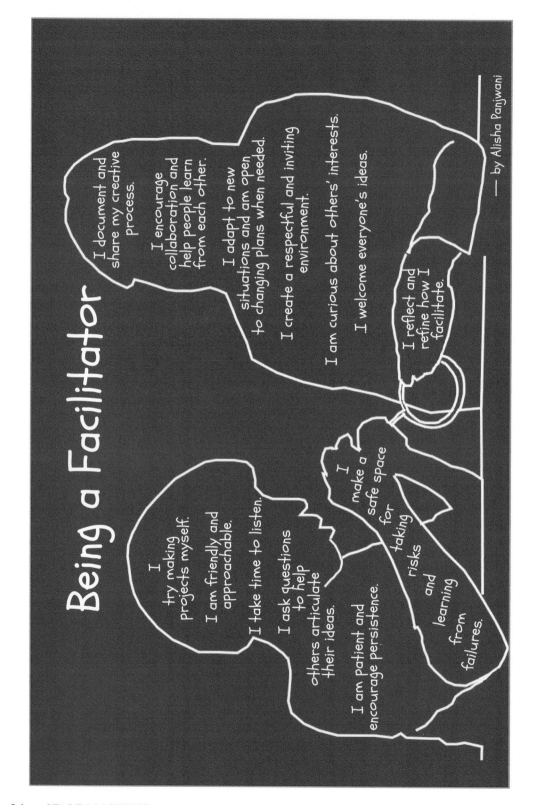

Being a Facilitator

I try making projects myself.

I am friendly and approachable.

I take time to listen.

I ask questions to help others articulate their ideas.

I am patient and encourage persistence.

I make a safe space for taking risks and learning from failures.

I document and share my creative process.

I encourage collaboration and help people learn from each other.

I adapt to new situations and am open to changing plans when needed.

I create a respectful and inviting environment.

I am curious about others' interests.

I welcome everyone's ideas.

I reflect and refine how I facilitate.

— by Alisha Panjwani

My Start Making! session wishlist...

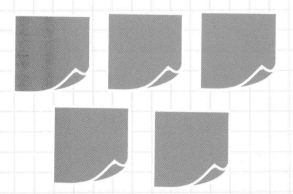

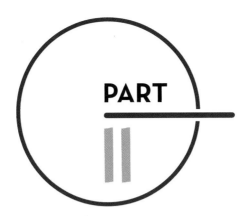

PART II

Start Making! Sessions

Start Making!
SESSION FLOW

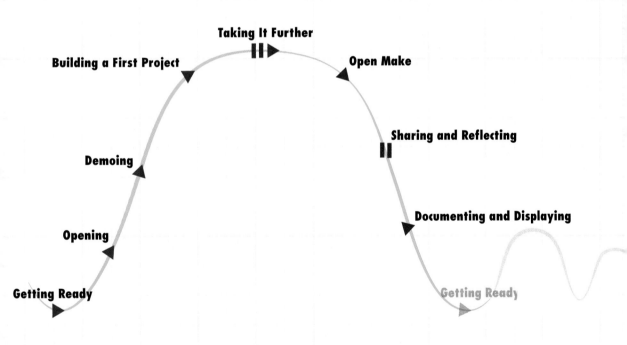

Taking It Further

Building a First Project

Open Make

Sharing and Reflecting

Demoing

Documenting and Displaying

Opening

Getting Ready

Getting Ready

Start Making! Session Flow

In this chapter we provide an overview of how each Start Making! session typically flows. We describe each aspect of the session flow—from getting ready for the session to documenting and displaying projects at the end. We encourage you to adapt this flow to meet the needs and interests of your makers.

SESSION GOALS

For each session, we provide a list of goals. The goals include gaining experience with new tools, exploring new concepts, and developing a range of skills.

✔ GETTING READY

Before each session, you will gather materials and set up the space. We provide a list of suggested materials for each session.

We also recommend that you create one or more projects of your own before the session. This will help you get to know the activity and create examples that you can share during the demo.

FACILITATION TIP

Organize the Space for Creative Exploration

Here are some tips for setting up a safe and inviting environment:

- Set up stations where makers can use tools that might get messy or that need facilitator support. For example, you can set up stations to help makers with hot gluing, soldering, or cutting with sharp instruments.
- Check to make sure that all the materials for the session are available and visible throughout the session.
- Create a box or bin for each of the Start Making! sessions. Keep the materials specific to each session together so that you can easily organize and find them when they are needed.
- Place sample projects on display to spark interest and ideas.

OPENING

Greet the youth and initiate an icebreaker or a warm-up activity to build relationships among the youth as well as the facilitators.

Opening activity (CASA DE LA JUVENTUD CLUBHOUSE, MORA, COSTA RICA)

Gather the youth into a circle around a table containing only the necessary materials. Share one or two example projects. Then, demonstrate how the project works and the basic steps makers need to take to get started designing their own project.

Introducing how to make a paper circuit (FLAGSHIP CLUBHOUSE, MUSEUM OF SCIENCE, BOSTON, MA)

Introducing an activity (CLT CLUBHOUSE, BANGALORE, INDIA)

FACILITATION TIP

Be a Maker First

Clubhouse Coordinator Yarelis Garcia was interested in offering Start Making! for the youth in her Clubhouse at the Boys & Girls Club of Metro West in Framingham, Massachusetts. She didn't have any experience with hands-on DIY (do-it-yourself) engineering projects. She decided to embrace becoming a maker herself before she tried out the activities with her young members and volunteer mentors.

She built her own projects as described in the Start Making! sessions, and then she displayed them on a shelf to spark interest before her Start Making! weekly program began. The signups for her program were overflowing. The youth appreciated Yarelis's enthusiastic embrace of learning and her willingness to present her first attempts proudly to everyone!

⇒ BUILDING A FIRST PROJECT

Get hands-on! Prompt makers to build their first version of the project by following the demonstrated steps and adding their own personal flair.

Making an initial project (YOUTH CONNECTIONS CLUBHOUSE, LISMORE, AUSTRALIA)

 ## TAKING IT FURTHER

Add new elements to the first basic activity, or prompt the makers to come up with their own versions to work on. Here you'll share more examples and help makers generate even more ideas.

 ⭐ **FACILITATION TIP**

Take a Break and Have a Snack!

Food and a break from the sometimes intense work of making are often necessary, and they provide a chance for everyone to relax, chat, and form bonds.

Experimenting with an activity (YOUTH CONNECTIONS CLUBHOUSE)

✂⚙ OPEN MAKE

Help the makers imagine, design, and make final projects, either individually or in small groups. Build on earlier activities and sessions to incorporate new interests, media, or tools.

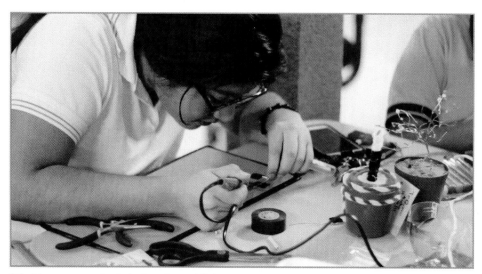

Working on a project during Open Make time (CASA DE LA JUVENTUD CLUBHOUSE)

OPEN MAKE BOX

In addition to the basic materials list for each activity, we also recommend you create an Open Make box that includes a broader range of materials and tools. These boxes can be used during the Open Make times to expand creative possibilities for making a variety of projects.

How to Build Your Open Make Box

These are our suggestions. You can replace or add your own items to the list.

- Butcher paper
- Foam sheets and boards
- Cardboard
- Felt sheets
- Mini hot glue gun and glue sticks
- Tacky glue
- Electrical tape and/or duct tape

- Scissors
- Precision knives and cardboard cutters
- Markers
- Glitter pens
- Ping-Pong balls
- Necklace cords

- Pin backs
- Ribbon
- Googly eyes, sequins, beads
- Embroidery thread
- Wooden craft sticks
- Other craft materials

Your Open Make box can also include leftovers or scraps from earlier sessions. Save packaging from other materials, including snacks, and encourage youth to scavenge around their homes for items such as egg cartons, plastic or foam food containers, empty plastic bottles and caps, and boxes.

Taking the Open Make Box Further

Depending on the technical skills of your makers and facilitators, you may also want to create a box with more complex or advanced tools. Here are just a few items you could collect in this box:

- Soldering irons
- Other conductive materials, such as conductive paint and/or pens
- Bandage scissors (perfect for cutting cardboard)

- Power drills

- Basic hand tools like hammers, saws, and box cutters

- A multimeter, to test your circuits; especially helpful for e-textile circuits

- Electronic prototyping tools, such as Arduino, Raspberry Pi, LilyPad, Chibitronics circuit stickers, or BBC micro:bit

- Safety goggles

★ FACILITATION TIP

Gathering Tools and Materials

You can improvise to pull together your own set of materials, depending on what is available in your area. Engage young makers, family members, and/or other community members to help you gather materials.

It may be difficult to find some of the electronic materials—such as copper tape, LilyTiny boards, and conductive thread—in local craft or hardware stores. You can find and order these materials online (see the Additional Resources list). Also, you can save money if you order in bulk and in advance.

A NOTE ON SAFETY

Provide support and coaching for new makers to use the tools successfully. Many facilitators in Clubhouses and other youth maker spaces advise developing your own safety procedures, using language and images that your young makers will quickly understand. It's not that they aren't capable; you just want to make sure they use the tools safely and efficiently and that they have fun!

We recommend safety goggles to protect against anything that might hit the eye. These are especially important when makers are taking apart devices, soldering, or working with electronics parts. Adults should wear safety goggles while doing these activities too, not just to keep themselves safe, but also to model safe practices for the youth.

We also recommend setting up specific stations for sharp instruments, glue guns, or other tools that may require adult supervision when they are being used.

SHARING AND REFLECTING

After the Open Make, encourage makers to come back together as a group to reflect on what they made together (or to spend some time reflecting in their journals), to share their work with others, and then to clean up together.

Sharing ideas (FLAGSHIP CLUBHOUSE)

CREATE AND USE JOURNALS

Individually, each maker can create or decorate a notebook in which they can brainstorm ideas, document projects, and also reflect on their process.

 Journal Prompts

Ask reflective questions and suggest that makers respond to them in their journals.

What was my favorite material used today?

What changes should I make or add to my project?

What are three things that I want to make using this technique?

★ FACILITATION TIP

Make Your Own Journal

You can use loose sheets of paper, a hole punch, a rubber band, and a paper-clip to make your own journal. Here are steps to make a quick and easy DIY journal:

1. Stack sheets of paper on top of each other.

2. Add construction or other thick paper on the top and bottom of the stack to form a cover.

3. Make two holes on the edge using a hole punch.

4. Loop a rubber band through the holes from the back of the paper.

5. Use a paperclip to hold the two ends of the rubber band together.

6. Add your name and decorate the cover!

The process of creating a journal is a great icebreaker and a way to encourage makers to document their own design process. You can encourage your makers to decorate their journal covers to make them their own!

Journals made by makers (FLAGSHIP CLUBHOUSE)

IDEA WALL

Set aside an area where makers can post ideas in a common place, such as on a whiteboard or a wall of the room, which they can draw from later. These ideas can include thoughts on materials, tools, techniques, stories, questions, or problems they want to solve.

📷 DOCUMENTING AND DISPLAYING

Encourage and help youth to document their process. They can draw sketches, take photos, and record video describing what they are making. Set up a physical space in which to display their creations. Make sure the display space is visible to everyone. You may also want to provide them with opportunities to post and share their projects and stories online.

Project display area (AWA CITY CLUBHOUSE, WHANGANUI, NEW ZEALAND)

CHOOSING ACTIVITIES AND ADAPTING SESSIONS

You can choose how many sessions and which parts of the suggested Session Flow to include in your own version of the Start Making! design, depending on how much time you have.

Here are a few hints on how to adapt each session to your situation:

Know your makers. If you're engaging youth and mentors who are new to your space, you'll need to set aside more time for icebreakers and other activities that help them get to know each other. Also, you may want to simplify the session to focus on just building an initial project; you can wait until later in the program to offer Open Make opportunities.

Find the buzz. Poll your young makers ahead of time to find out what materials, technology, and projects excite them most.

Keep an open mind. Makers might be interested in any of the materials and tools, so keep an open mind and provide a variety of options rather than making assumptions based on gender, age, or background. Encourage everyone to participate in all of the session activities, such as sewing, electronics, and robotics.

Try something new. You can find a wealth of ideas for creative maker projects online that could fit well within the Start Making! program design. Invite everyone to suggest activities or projects based on their own interest areas.

ADDITIONAL RESOURCES

▶ Start Making! demo videos for facilitators:
  bit.ly/start-making-facilitation-playlist

▶ Online stores for electronic materials:
  adafruit.com, sparkfun.com, digikey.com

▶ Maker Shed online store:
  makershed.com

 FACILITATION TIP

Hack This Guide

You'll notice that we've included some reflection pages within this guide on which you can write or sketch your ideas, plans, and questions. We encourage you to add notes and reflections throughout the guide—and make it your own!

What will I add to my Open Make box?

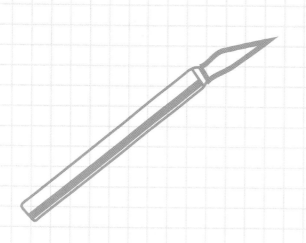

Light It Up:
PAPER CIRCUITS

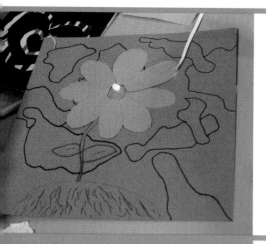

New makers will make paper creations that light up in interesting and surprising ways. In the process, they will learn how to make electrical circuits.

SESSION GOALS

In this session, makers will

- Build a circuit on paper using conductive tape, LEDs (light-emitting diodes), and coin cell batteries.

- Design a light-up project, such as a greeting card.

- Learn about basic electrical circuits and explore ways to create projects with circuits.

✔ GETTING READY

You're about to begin your group's journey with Start Making! As a first step, work with others to set up your space so it is welcoming and interesting. Arrange materials and tools so they're organized and accessible, while keeping safety in mind.

Before the session, make your own paper circuit project using the materials and instructions that follow. We recommend making at least two examples to help participants imagine a range of possibilities. If you have other facilitators working with you, you can each work on making an example to share.

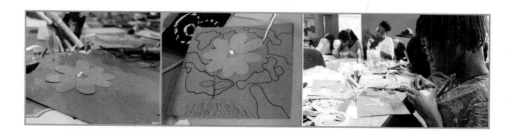

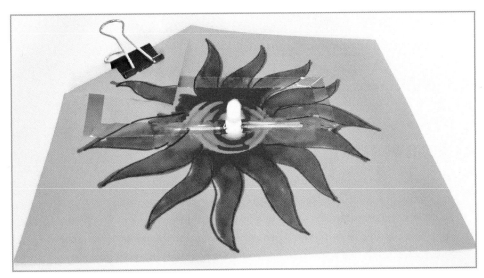

Example of a paper circuit

MATERIALS

- ▶ LED lights (one or more colors)
- ▶ 3V coin cell batteries (such as the CR2032)
- ▶ Copper tape (conductive foil)
- ▶ Paper (note cards, cardstock, construction paper, paper cups, or other paper)
- ▶ Scissors
- ▶ Clear tape
- ▶ Binder clips (small, about ¾-inch wide)
- ▶ Colored pens, pencils, or markers

SPACE AND TOOLS SETUP

Lay out the materials on a table that will be easy for everyone to reach.
Group the tools, such as scissors, together. Make sure there is enough open
table space for participants to work on their projects.

CIRCUITS

All electronics—from computers and phones to the lights in your home—
use circuits to conduct electricity. All circuits include a power source (such
as a battery) connected by wires to something that requires power (like a
light bulb). In this activity, you'll use copper tape for the wires, a round coin
battery to supply the power, and a small LED to function as a light bulb.

OPENING

To begin your first session, start with an icebreaker to help everyone get
to know each other. Gather in a group and have each person share their
name, where they're from, and something they like to make (or their favorite
activity).

DEMOING

Share two or three different examples of paper circuits with the group. Pass
around the examples and briefly describe your process. Encourage members
of the group to imagine what kind of project they might want to make.

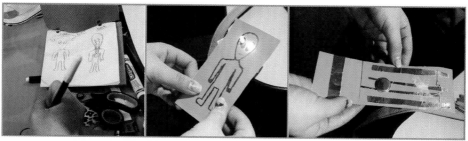

Facilitator sharing how she made a light-up paper circuit

BUILDING A FIRST PROJECT

For their first project, makers explore how to make a simple circuit that lights up an LED on a piece of paper. This can provide the foundation for many related projects.

Make the LED light up.   Give one LED light and a battery to each maker. Encourage them to explore the look and feel of these materials and share what they notice. Then, ask them to try to make the LED light up.

 FACILITATION TIP

Encourage Experimentation

You may want to try starting with very little instruction. Allow the makers to jump in and try to make it work. Moments of "failure"—when things aren't working as expected—can present you with opportunities to encourage persistence and allow makers to practice problem solving.

Lighting up an LED with a battery (SCI-BONO CLUBHOUSE, JOHANNESBURG, SOUTH AFRICA)

Notice that the battery has a symbol on each side. One side has a plus (+) sign, which stands for the positive terminal. The other side has a minus (-) sign, which stands for the negative terminal.

The LED also has positive and negative terminals. The longer leg is positive and the shorter one is negative.

To make the LED light up, connect the negative side of the battery to the negative leg of the LED and the positive side of the battery to the positive leg of the LED.

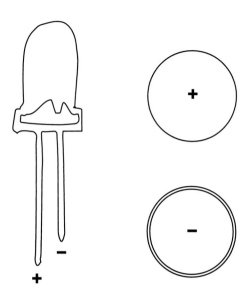

Introduce the circuit layout. Once everyone is able to make their LED light up, introduce this basic layout that they can use to make their circuit on paper. You can hand out printed copies, or you can draw the diagram on a board for everyone to refer to when they are making their own.

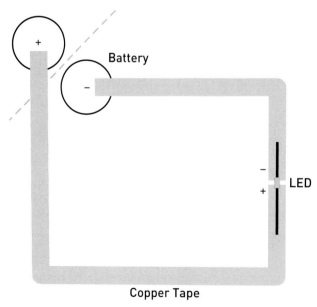

Basic layout for a paper circuit (DESIGNED BY JIE QI)

Below are the steps for making the paper circuits, which you can use to help guide the makers.

1. **Add copper tape.**

   Take a piece of paper and on it, place copper tape so it follows the lines in the diagram.

   To make turns, tape until you reach the corner where you want to turn. Then, fold the tape to turn the corner.

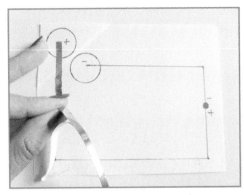

Start placing the copper tape in the center of the battery circle to make sure that it will make contact with the battery.

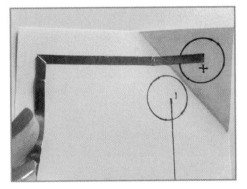

Folding the copper tape to turn corners

2. **Attach the LED.**

Take the LED and spread the legs so that they are sticking out to each side. Place the LED on top of the copper tape, so the legs touch the tape. Then secure them in place with clear tape.

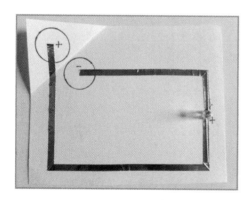 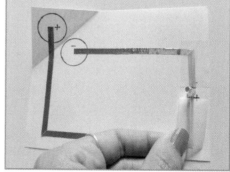

3. Add the battery.

Place the battery negative (–) side down, where the circle with the (–) sign is. It should touch the copper tape. Next, fold over the corner of the paper so that the tape going to the positive (+) circle touches the battery. Your light should turn on.

4. Complete the circuit.

Use a binder clip to hold the circuit in place. If the light doesn't turn on, check that the copper tape is running to both sides of the battery and the LED with no breaks, and that the two tape lines don't touch each other.

FACILITATION TIP

Create Opportunities for Peer Mentoring

You can encourage new makers to learn from each other. Point out when you notice someone trying a new technique and ask questions to help others learn from their process. When someone has figured out how to solve a problem, you can suggest that they offer to help others. Encourage them to help the other person learn the process, rather than do the work for them.

Peer mentor helping to make a circuit (HENNEPIN COUNTY LIBRARY BEST BUY TEEN TECH CENTER, MINNEAPOLIS, MN)

TAKING IT FURTHER

You can show examples of other types of circuits, such as a circuit with a switch or more than one light.

To add a switch, make a break in the circuit by removing a piece of the copper tape. Then, add a way to connect the two parts with something conductive, such as by putting a piece of foil on your finger or by adding a flap of paper with copper tape on it. (For more instructions, see the paper circuit basics tutorial in the list of resources at the end of this chapter.)

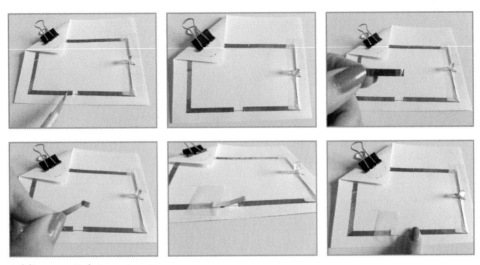

Adding a switch to a circuit

If you want to add more lights to a circuit, you can connect them in two different ways:

Parallel circuit To make a parallel circuit, add the lights so that each has its own row. Make sure to connect the positive ends of all the LEDs to the positive side of the battery.

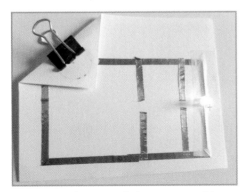
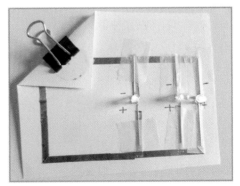

You can make a parallel circuit to add more than one light.

Series circuit To make a series circuit, place the lights in a single loop, connecting the positive end of one LED to the negative end of the other. For a series circuit, you need more batteries to power more lights—one for each light.

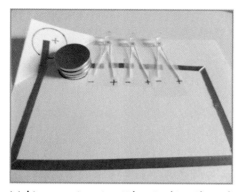 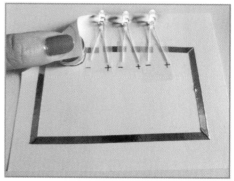

Making a series circuit by stacking three batteries to power three lights

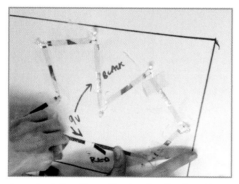 

A series circuit on the back of a drawing (with the person's hands connecting across the break in the tape)

Light-up project inspired by a favorite memory (CREDIT: ALISHA PANJWANI)

GENERATING IDEAS

After making a simple circuit, gather the group to discuss what they noticed and to brainstorm ideas for what they might want to make with paper circuits using what they've learned so far. Have them jot down their ideas and post them on the idea wall.

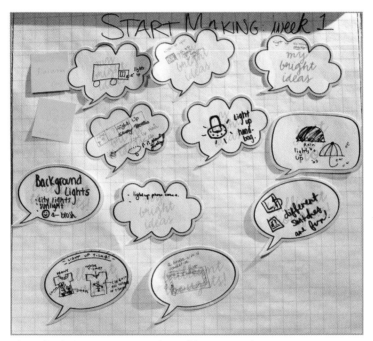

Ideas for light-up projects shared by new makers

✂⚙ OPEN MAKE

Provide time for the makers to experiment and design their own projects. They can pursue ideas they've shared on the idea wall or try some of the following ideas.

MAKE A DRAWING

Make a drawing that lights up. Poke a small hole in the paper where you want the LED light. Push the top of the LED through the hole and bend the legs. Connect the tape and battery on the back side of the paper.

A drawing with lights added

LIGHT-UP ORIGAMI

Make an origami sculpture and add a paper circuit to make it light up. (See the paper lantern tutorial in the resource list at the end of the chapter for ideas.)

Light-up origami examples (YOUTH CONNECTIONS CLUBHOUSE, LISMORE, AUSTRALIA [LEFT]; NEVE YOSEF CLUBHOUSE, HAIFA, ISRAEL [RIGHT])

ILLUMINATE YOUR JOURNAL

Add paper circuits to your journal. Make a light-up portrait of yourself or your favorite activity.

 ## SHARING AND REFLECTING

Give the makers time to reflect and note their thoughts and ideas in their journals. Then gather everyone together and have them share their projects and experiences with the group.

 # Journal Prompts

You can ask the following:

What was my favorite material from today's activity?

What are some things that I want to change or add to my paper circuit?

What are three more things that I want to make using this technique?

📷 DOCUMENTING AND DISPLAYING

Invite makers to display their projects on a shelf, wall, or online gallery. Encourage them to take photos and document their experiences to share the process with other makers. They can also include drawings of the circuits they created so that they remember how to build circuits for future projects.

A shelf displaying light-up creations by new makers (FLAGSHIP CLUBHOUSE, MUSEUM OF SCIENCE, BOSTON, MA)

STORY FROM THE CLUBHOUSE NETWORK: AN EXAMPLE OF A FIRST SESSION OF START MAKING!

To give you a sense of what a Start Making! session is like, here's a description of the first session at one of the Clubhouses.

A group of 11 girls showed up for the first day of the Start Making! program, which was being offered on Saturday mornings. During the opening icebreaker, the girls introduced themselves and shared something they liked to do. A couple of the girls were shy and reluctant to talk, whereas several others were lively and talkative. The facilitators strived to make everyone feel welcome and comfortable. Some of the girls expressed interest in learning to make things; others mentioned that they came to spend time with their friends.

The facilitators shared a few examples of light-up cards, which they had made before the new makers had arrived. They passed the examples around the table, and the girls looked carefully at the cards and seemed curious about how they could make their own.

The facilitators gave each girl a battery and an LED and encouraged them to figure out how to make it light up. Some of the girls figured out how to make it work on their own, while others needed help. Getting their lights glowing generated excitement among the girls and heightened their interest in doing more.

The facilitators then suggested that the girls make personalized maker journals to document what they were making. They provided precut construction and printer paper, paper clips, pens, and other materials. The process of making and personalizing the journals gave the facilitators and new makers a chance to talk and get to know each other.

Examples of handmade journals created by new makers (FLAGSHIP CLUBHOUSE)

The facilitators then brought out the copper tape and other materials so the girls could start making paper circuits. The facilitators shared the basic template for a circuit, which one of the facilitators had posted on the wall. The girls drew the circuit on a piece of paper and started using copper tape to make their circuits.

The facilitators encouraged those who had their circuits working to help the others who were encountering difficulties. The most common problem was that the connections from the copper tape to the battery or LED were loose. Also, some had cut the copper tape instead of folding it at the corner, so the circuits were not fully connected. In the end, everyone was happy to get their circuits working.

During the open make time, the girls made a range of personalized light-up projects. Several made flowers or hearts with LEDs in the center. Some made projects that reflected activities they liked to do. For example, one girl made a guitar that lit up. Other projects included a person with bright eyes, a clown with a light-up nose, and a glowing jellyfish.

For the closing, everyone came together in a circle to share their projects. The facilitators asked the girls questions about their experiences, using the reflection prompts, and then encouraged the girls to write or draw their reflections in their journals. Those who had been reluctant to speak at the beginning of the session expressed themselves in other ways, by sharing their projects or by writing their reflections in their journals.

After the girls headed home, the facilitators reflected on how the first session went, sharing what went well, what they had questions about, and what they thought could be improved.

They knew that the first session sets the tone for the rest of program, so they felt it was worth all the planning and preparation to help the new makers feel welcome to start making together.

DESCRIBING CIRCUITS

Electricity flows in a **circuit**.

Electricity is the flow of particles called **electrons**.

The **circuit** is a path that allows electrons to flow in a loop: from the battery, through the LED, and back to the battery.

This stream or flow of electrons is called **current**. The current causes the LED light to turn on. The current will not flow if the circuit is not complete.

An **LED** (light-emitting diode) is a type of light that allows the current to flow only in one direction. This is the reason why the positive leg of the LED needs to connect to the positive side of the battery, and the negative leg of the LED to the negative side of the battery. Otherwise, the LED will not light up.

A material that allows the current to flow easily is called a **conductor**. All metals are good conductors of electricity. We used **copper tape** with sticky adhesive on the back to conduct electricity from the battery to the LED light.

Insulators are materials that do not allow the current to flow easily. Paper or clear tape can act as a good insulator to prevent pieces of wire or conductive tape from touching each other.

A **switch** is used to complete a circuit or break a circuit and hence can be used to make the lights turn on and off.

There is only one path for the current to flow in a **series circuit**. The current will go through one LED, then through the next LED, and so on.

There are two or more paths for the current to flow in a **parallel circuit**. The current flowing through one LED is separate from the current flowing through the other LEDs.

STORY FROM THE CLUBHOUSE NETWORK: MAKING LIGHT-UP BADGES AND BUTTONS

Facilitators who led Start Making! at their Clubhouses found that new makers enjoyed the paper circuit activity and often applied the skills they'd learned to make other light-up projects. For example, Cindy Priester, Clubhouse

Coordinator at the Southeast & Armed Services YMCA in Colorado Springs, CO, suggested that the new makers in her Clubhouse create name badges to introduce themselves when new people joined their program. They made light-up name badges using note cards with a pin on the back. They expressed their personality through their designs, adding LEDs to light up their names and images of themselves.

She explained how they later built on this idea:

Makers at our Clubhouse had the idea to take their first paper circuit name badges further to create buttons with words of encouragement (such as "Keep trying!") or to describe a maker "superpower" (such as "creativity"). Later, when the circuit-based projects got more challenging, I would sometimes suggest that they make a quick paper circuit note card or badge so that they could practice their understanding of basic electronics and get a quick success to remind them they could do it. Later, they demoed their perseverance when they participated in Maker Media's Maker Camp and posted their light-up button innovations to the online community.

ADDITIONAL RESOURCES

▶ Paper circuit basics:
hlt.media.mit.edu/?p=2505

▶ Flapping origami crane:
hlt.media.mit.edu/?p=1448

▶ Paper lantern tutorial:
evilmadscientist.com/2008/paper-circuitry-at-home-electric-origami/

Start Making! makes me curious about...

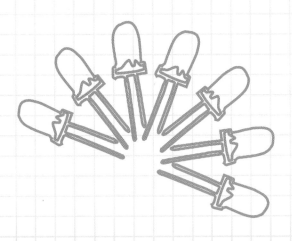

Make It Sing:

PROGRAMMABLE MUSICAL CREATIONS

New makers will create musical projects using everyday materials connected to a computer with a MaKey MaKey board. In the process, they will experiment to figure out which materials are conductive and learn to code a simple computer program using the Scratch programming environment.

SESSION GOALS

In this session, makers will

- Build music-making projects using a pencil sketch and other conductive materials connected to a MaKey MaKey board.

- Get started with computer coding by remixing a Scratch musical project.

- Explore and become curious about the electrical properties of everyday materials.

✔ GETTING READY

Get ready for a playful session experimenting with sounds and music!

For this session, you'll need the MaKey MaKey invention kit. You'll also need a computer running Scratch, a free creative programming environment. Watch the introductory videos for MaKey MaKey and Scratch, and try exploring the "getting started" guides for each of the tools.

Build your own musical drawing using the steps described for the first project.

Looking at a MaKey MaKey (CASA DE LA JUVENTUD CLUBHOUSE, MORA, COSTA RICA)

Making arrows to connect to a MaKey MaKey (BOYS & GIRLS CLUBS OF METRO WEST CLUBHOUSE, FRAMINGHAM, MA)

MAKEY MAKEY INVENTION KIT

With MaKey MaKey, you can use everyday objects to create your own physical interfaces to the computer. For example, you can make a piano keyboard out of pieces of fruit or use Play-Doh to create a controller for a game. You plug the MaKey MaKey board into a computer and then attach a banana (or other conductive object) to the MaKey MaKey. When you touch the banana and complete a circuit on the MaKey MaKey, the computer responds as if a key has been pressed. To see the introductory video and find out more, visit makeymakey.com.

Pressing a key using a MaKey MaKey board with Scratch

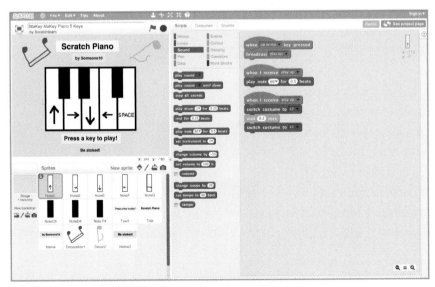

A Scratch piano project to play with the MaKey MaKey

SCRATCH

Scratch is a programming environment designed for young people (ages 8 and up) to create animations, games, stories, and other interactive projects. To code in Scratch, you snap together colorful blocks on the screen and combine them to program images, text, music, and sounds. You can also share and remix projects in the Scratch online community. Scratch is available free of charge and has been translated into more than 40 languages. Scratch is developed by the Lifelong Kindergarten group at the MIT Media Lab. To start creating, visit scratch.mit.edu.

MATERIALS

▶ MaKey MaKey kits, which include a circuit board, alligator clips, and a USB cable, are available at makeymakey.com.

▶ Computers with Scratch. You can use the online version of Scratch at scratch.mit.edu or download and install the offline editor from scratch.mit.edu/scratch2download.

▶ Scratch piano project: scratch.mit.edu/makeypiano

▶ Clipboards with metal clips

▶ Pencils, #2 or softer (graphite, not color)

▶ Blank paper

▶ Conductive materials, such as bananas, apple slices, gum drops, or other moist foods; Play-Doh or other water-based clays; aluminum foil, spoons, copper tape, or other metal objects

▶ Non-conductive materials, such as plastic, electrical tape, and cardboard

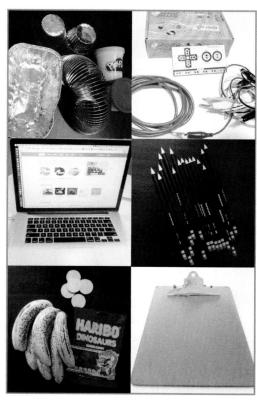

SPACE AND TOOLS SETUP

Arrange the materials. Lay out paper, pencils, and clipboards on a table for the first activity. In addition, for the Open Make time, set up a side table with a variety of conductive and nonconductive materials.

Set up the computer stations. Plug the MaKey MaKey board into each computer using the USB cable.

Open Scratch with the MaKey MaKey piano project running: scratch.mit.edu/makeypiano. (If you plan to use the offline version of Scratch, download the project and open a copy on each computer.)

For the demo computer, attach alligator clips to each of the MaKey MaKey arrow key ports (up, down, left, right) and one to the Space key port. In addition, attach one alligator clip to one of the ports labeled Earth. (You may want to use the black wire for Earth so that you can remember which wire is ground.)

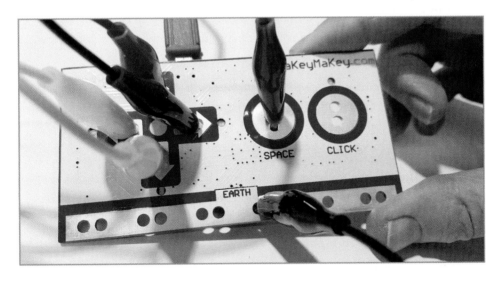

🚪 OPENING

You may want to begin with an icebreaker or a group activity to help makers get to know each other better. For example, in the spirit of this session's focus on music, you could have each person in the circle act out playing a musical instrument, and have the others guess what instrument it is and then pretend to play it themselves.

 DEMOING

Introduce the MaKey MaKey to spark interest and show the makers how they can make their own musical drawings.

Show a musical drawing. Demonstrate a simple musical drawing project connected to MaKey MaKey and Scratch. Show how you can play it, and ask one of the makers to try playing it, too.

Point out how you need to ground yourself by touching the metal clip that attaches to the Earth port on the MaKey MaKey. Then touch the thick pencil lines in order to complete the circuit and make it play.

Complete the circuit with just your body. Show how you can create a circuit with the MaKey MaKey by using your body to conduct electricity. Hold the metal part of the clip attached to Earth on the MaKey MaKey board in one hand. With your other hand, touch the metal part of another clip. You should hear a different piano note each time you complete the circuit with your body depending on which clip you touch: up arrow, down arrow, left arrow, right arrow, and Space key.

Complete the circuit with more than one person. You can also have the makers try activating the MaKey MaKey together by holding hands. Ask one person to touch the Earth port and another person to touch the Space key port. When they touch hands the circuit should play.

Ask the makers these questions: Why does this work when I just use my hands? Am I conductive? What other objects are conductive? What makes them conductive?

You may want to raise these questions now, and then discuss them at the end of the session, after the makers get a chance to experiment and explore different materials.

Completing a circuit by holding alligator clips

BUILDING A FIRST PROJECT

For their first project, makers can sketch and then play a musical drawing.

This activity was developed by Jay Silver, one of the creators of the MaKey MaKey, and was featured in Intel's *Sketch It, Play It* guide. For inspiration, you can show a video of the activity, available at start-making.org.

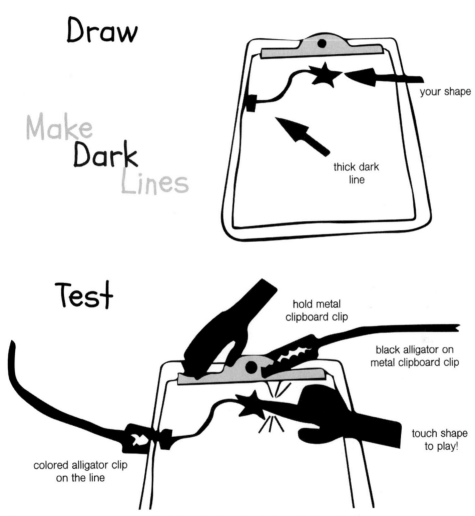

How to make a simple musical drawing to attach to a MaKey MaKey

Here are steps for making a musical drawing that connects to the Scratch piano project.

1. **Draw.**

   a. Insert a piece of paper into the clipboard.

   b. Use a pencil to draw a star or other solid shape on the paper.

   c. Draw a thick, dark line from the shape to the paper's edge.

2. **Connect.**

   a. Attach a wire from Earth on the MaKey MaKey to the metal clip on the clipboard.

   b. Clip one of the wires from a MaKey MaKey port, such as the up arrow, to the end of the line you drew.

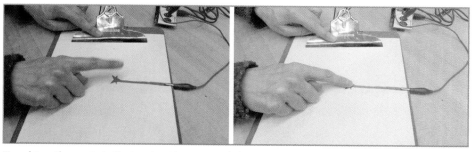

Touching the metal clip to play the musical drawing

3. **Test.**

   a. Touch the metal clip with one hand, then touch the shape to hear the note play.

   If the note doesn't play, see the Troubleshooting Tips on page 66.

4. Add more.

a. Now draw more shapes so that you can attach each shape to one of the MaKey MaKey ports (the up, down, left, and right, arrows and Space).

OPEN AND CLOSED CIRCUITS

You can use the MaKey MaKey to demonstrate the difference between open and closed circuits. To activate a key on the MaKey MaKey, you need to make a closed circuit.

Open Circuit A circuit with a path that has a break in it so that electricity does not flow

Closed Circuit A circuit with a path that is connected, allowing electricity to flow

If you only touch a wire going to Earth on the MaKey MaKey, you have an open circuit.

If you also touch a wire going to a MaKey MaKey key port, you have a closed circuit, with the electricity flowing through you. The computer will think you pressed a key!

5. Play.

Now you can play your musical drawing.

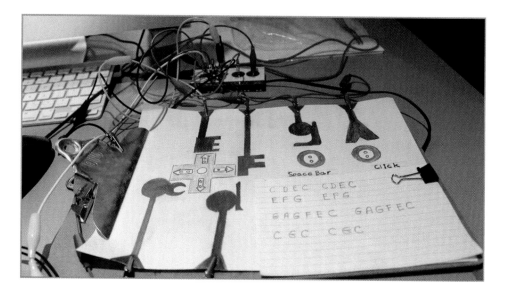

TAKING IT FURTHER

Introduce the concept of conductive materials by demonstrating how a banana, a foil pan, or modeling clay can be connected to the MaKey MaKey board; then touch the conductive object (while also touching the Earth port) to trigger a sound in the Scratch program.

Table laid out with conductive and nonconductive materials (FLAGSHIP CLUBHOUSE, MUSEUM OF SCIENCE, BOSTON, MA)

TERMS TO KNOW

Conductivity Conductivity refers to the flow of electrons in a material.

Conductors Some metals, such as copper, conduct electricity better than others. Pencils have enough graphite (a semimetal) in them to work as a conductor, so you can "draw" a conductive path that allows electrons to flow to the MaKey MaKey. Examples of materials that also have conductive properties include moist fruit, modeling clay, and even the human body.

Insulators On the other hand, some materials keep their electrons close and tight. Plastic, rubber, fabric, glass, and wood don't conduct electricity. These materials make good insulators.

Suggest that makers work in pairs to experiment with using conductive (and nonconductive) materials. They can disconnect their musical drawings and experiment with using other materials to trigger the sounds on the Scratch piano.

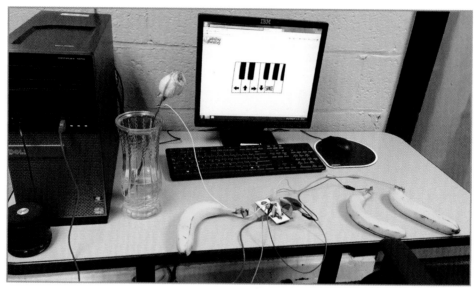

Bananas and a flower interface to a Scratch piano project

OPEN MAKE

Now makers can expand their musical projects.

To spark ideas, play with other materials that can trigger keys on the computer. Make a bigger instrument out of a foam board and other materials, such as spoons! For inspiration, watch the MaKey MaKey Music Examples video created by Eric Rosenbaum (bit.ly/makeymusicvideo).

Using conductive materials to connect to a MaKey MaKey (YOUTH CONNECTIONS CLUBHOUSE, LISMORE, AUSTRALIA)

Circulate among the groups and prompt the makers to keep experimenting, building, and taking their play in new directions. Encourage ways for more than one person to interact with the musical project.

★ **FACILITATION TIP**

Find Conductive Materials

Playing with MaKey MaKey boards is a fun way to test the conductivity of materials in the world around you. Try funny or unexpected objects, materials, and foods to use as conductors.

Mentors from the Flagship Clubhouse at the Museum of Science in Boston encouraged their makers to test the conductivity of canned spray cheese. They created a circuit conducting electricity all the way around the table!

HACK YOUR MUSICAL INSTRUMENT PROGRAM

Suggest editing and remixing the Scratch piano to play different sounds. Or, try using the Scratch MaKey MaKey drum project: scratch.mit.edu/makeydrum.

A MaKey MaKey board can control anything on the computer that is triggered by the keyboard. What other software programs move or make noise on the computer when you press a key?

Using Play-Doh to connect to a MaKey MaKey (FLAGSHIP CLUBHOUSE)

MAKING A COMPUTER PROGRAM

A computer program is a set of instructions that tells the computer what to do. You can write your own computer program using a programming language, such as Scratch. To program, you think about what you want to happen and then break it into smaller steps in a language the computer can understand. In this session, you create or remix a program that tells the computer what note to play whenever a key is pressed.

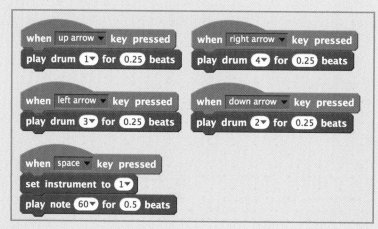

An example of a Scratch program that plays different sounds when keys are pressed

Experimenting with conductive materials (SCI-BONO CLUBHOUSE, JOHANNESBURG, SOUTH AFRICA)

Here are some of the Open Make projects created by Clubhouse makers. *Do you see anything you want to try?*

Arrow key interface made with copper tape (YOUTH CONNECTIONS CLUBHOUSE)

Learn How to Play Piano project (BOYS & GIRLS CLUBS OF METRO WEST CLUBHOUSE)

Playing a dance game made using a MaKey MaKey (CEDES CLUBHOUSE IN ALAJUELITA, SAN JOSÉ, COSTA RICA)

Ice-Cube Keyboard (SORENSON UNITY CENTER CLUBHOUSE, SALT LAKE CITY, UT)

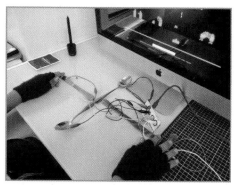

Interface made with spoons (YOUTH CONNECTIONS CLUBHOUSE)

SHARING AND REFLECTING

Provide time for the makers to share what they noticed and made during their explorations. Ask them to jot down their reflections in their journals.

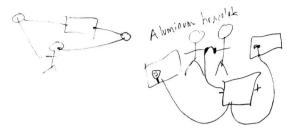

Sketch of ideas for connecting to MaKey MaKey

 # Journal Prompts

You can ask the following:

What was the most surprising conductive material you hooked up to your MaKey MaKey board? What else would you like to try?

How would you explain how your project works? Draw a diagram to show how it connects.

Who helped you with your project? What did they do to help?

📷 DOCUMENTING AND DISPLAYING

Don't forget to document these projects as you go along! Some projects may literally get eaten or dry up, but reflections can live on through photos or videos, online posts, or journal entries.

What name do you want to give to your musical project? What instructions will you provide for playing it?

You can also help the makers save and present their musical projects during the final Show & Share sessions. Friends and family members love playing with MaKey MaKey projects!

STORY FROM THE CLUBHOUSE NETWORK: THE CITY THAT SPEAKS

The makers at the East Palo Alto Boys & Girls Club Clubhouse in California used their MaKey MaKey boards not only to learn about conductivity and play with musical interfaces and coding, but to use them as tools for storytelling and community-building.

They started by creating a LEGO model of the White House and other famous places around Washington, D.C. The Clubhouse members then embedded a MaKey MaKey board in the city diorama to make parts of it interactive. They used Scratch to record and import sounds. The sounds included their own voices, lines from President Obama's speeches, police sirens, and even Sunny, the White House puppy, barking. They called the project The City That Speaks.

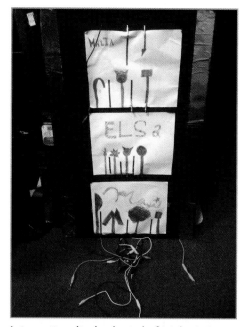

Interactive display board of Make It Sing projects for Clubhouse community showcase (EAST PALO ALTO BOYS & GIRLS CLUB CLUBHOUSE, EAST PALO ALTO, CA)

The City That Speaks made its way to Maker Faires in the San Francisco Bay Area and Washington, D.C., where Faire goers were encouraged to help build more of the city.

The City That Speaks model (EAST PALO ALTO BOYS & GIRLS CLUB CLUBHOUSE)

Adding on to the city model at the Bay Area Maker Faire

ADDITIONAL RESOURCES

▶ MaKey MaKey group activity guides:
makeymakey.com/guides/

▶ MaKey MaKey video gallery:
makeymakey.com/gallery

▶ Scratch learning resources and support:
scratch.mit.edu/help

Three things that I found the most surprising in this session are...

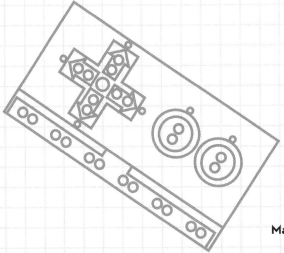

Paint with Light:
ILLUMINATED WANDS
AND PHOTOGRAPHY

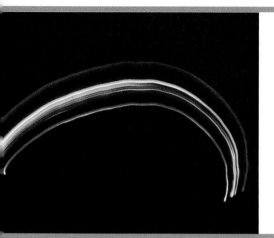

Makers will create simple LED wands and then use them to make light paintings, capturing images with a digital camera (with a long-exposure setting). In the process, they will learn about simple circuits and long-exposure photography.

✔ GETTING READY

This session has two parts: 1) building light wands and 2) creating light paintings.

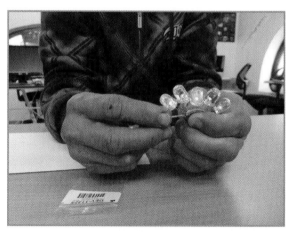

Experimenting with LED lights and a coin battery
(YOUTH CONNECTIONS CLUBHOUSE, LISMORE, AUSTRALIA)

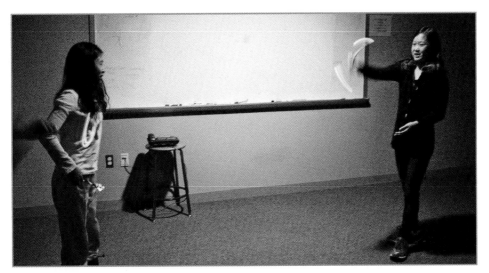

Waving LED lights to create a pattern (FLAGSHIP CLUBHOUSE, MUSEUM OF SCIENCE, BOSTON, MA)

To prepare for the session, try light painting yourself, ideally with at least one other person participating and helping with the camera. Save the photos from your light experiments, and make a slideshow of the images you create to share with your makers as examples during the demonstration.

This session is inspired by the Exploratorium's *The Art of Tinkering* by Karen Wilkinson and Mike Petrich (2014) and the work of the staff and mentors of the Flagship Clubhouse at the Museum of Science in Boston, MA.

MATERIALS

▶ Digital cameras with long-exposure settings, or smart phones with an app installed that allows for long-exposure photography (such as Slow Shutter Cam or LongExpo)

▶ Tripod (if available)

▶ Single-color LED lights

▶ Blinking RGB (tricolor) LED lights

▶ CR2032 (coin) batteries

▶ Pipe cleaners

▶ Clear tape

▶ Straws or craft sticks

▶ Translucent materials, such as wax paper, glass jars, ping-pong balls, drinking cups, and glue sticks

▶ Conductive materials to extend the circuit, such as copper tape or wire

▶ Additional light sources such as flashlights, toys with moving lights, cell phone screens, or electroluminescent (EL) wire (optional)

SPACE AND TOOLS SETUP

Set up a light and a dark space. Ideally, this activity uses two rooms or spaces: 1) a well-lit space for building light wands, and 2) a darkened room for taking photos of the lights in motion.

If you don't have two separate spaces, you can start by making the wands and then darken the room to take photos.

In the well-lit room, arrange the craft and electronic supplies for building wands.

In the darkened room, set up one or more digital cameras on a tripod or table.

Adjust the camera settings. Adjust the camera settings so that the exposure remains open. The camera should remain steady and not move, so you will need to set the long exposure setting (sometimes labeled "shutter priority" or "bulb") to longer settings (between 2 and 10 seconds).

Start by turning off the flash. Later, you can get some interesting effects if you turn the flash on again to get a pop of light to capture images of people.

TROUBLESHOOTING TIPS

▶ Take a few experimental shots and then adjust settings as needed.

▶ To keep the camera steady, use a tripod and a remote trigger or a self-timer.

▶ Explore advanced camera settings (such as high aperture, low ISO, manual focus, RAW format, no image stabilizer, and bulb mode).

OPENING

You can start the session with a group icebreaker. For example, during such a session at the Flagship Clubhouse at the Museum of Science in Boston, volunteer mentor Lauren led the creative activity called "Exquisite Corpse."

Each person was given a piece of paper and asked to fold it in thirds, from top to bottom. Each person then drew the head of a creature at the top of the sheet, without showing it to the person next to them. Then they folded it back so it was not visible and passed it to the next person. The next person added the creature's body without looking at what the last artist drew, folded it, and then passed it to the next person who drew the legs. When the sheets were unfolded, the makers were intrigued and amused by the drawings they had created together.

Opening activity (FLAGSHIP CLUBHOUSE)

⚡ DEMOING

Light painting is not an obvious process, since the light is captured over time. Make sure to begin by sharing examples to inspire and illustrate the activity.

1. **Show examples of light painting photos.**

   Introduce the idea of making light paintings. Show examples of light painting photos that you have made or found online. Ask the group what they notice about the different photos.

Introducing light painting activity (NEVE YOSEFY CLUBHOUSE)

2. **Share examples of light wands.**

   Show the makers examples of light wands that they will be able to make using LED lights and batteries.

LONG-EXPOSURE PHOTOGRAPHY

To make a light painting, you leave the camera's shutter open for a few seconds or more. This is known as *long-exposure photography*. By leaving the shutter open, you can capture the movement of light over time.

Exposure refers to the amount of light that comes into the camera for a photograph. The exposure is determined by how long the shutter is open, how wide the lens opening (*aperture*) is, and how much light is in the space.

---

3. Demonstrate how to capture light painting images.

You may also want to demonstrate the technique for light painting in the darkened room, including how to use the cameras for long-exposure photos.

Looking through a camera lens (NEVE
YOSEF CLUBHOUSE, HAIFA, ISRAEL)

Introducing light painting activity (NEVE
YOSEF CLUBHOUSE)

⇨ BUILDING A FIRST PROJECT

Begin by building simple LED light wands.

1. Make a simple LED wand.

Choose an LED. Use clear tape to attach the LED to a coin battery. Attach the longer leg of the LED to the positive side of the battery, and the shorter leg to the negative side.

Making light wands (FLAGSHIP CLUBHOUSE)

The LED should light up. (If the battery is flipped, it won't light up.)

Wrap the battery and legs in tape so that they are secure and none of the metal is exposed.

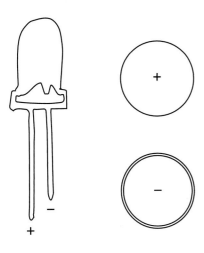

Add a pipe cleaner as a handle for your light wand. Wrap the pipe cleaner around the battery, and extend it so you have a long wand to hold the light.

(Because pipe cleaners have metal inside, make sure that the pipe cleaner only touches the tape and not the metal battery or LED legs.)

Wrap the pipe cleaner firmly so the light doesn't fly off when you wave it. You can add tape to make it more secure.

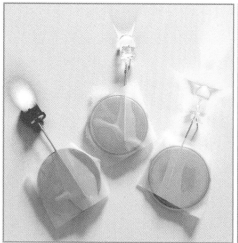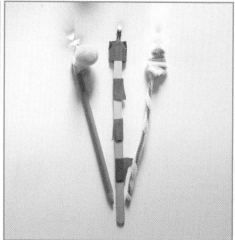

You can also use other materials to make wands, such as craft sticks, pencils, or straws.

You can also decorate your wand with craft materials.

2. **Try light painting.**

   After making the light wands, work in teams to experiment with light painting in the dark space.

   Arrange for someone to manage the camera and take the photos.

   Take turns creating light paintings individually or in small groups.

   Have the photographer press the shutter release button while the painter draws simple shapes in front of the camera.

Makers can alternate between building light wands and painting with light.

Making a light painting (FLAGSHIP CLUBHOUSE)

 ## TAKING IT FURTHER

Now that makers understand the basics of light painting, they can begin to explore more techniques and effects.

EXPLORING LIGHT EFFECTS

Show the group examples of light paintings that use more advanced effects and techniques, such as drawing silhouettes, using multiple colors, or writing a word.

Light painting made by drawing a silhouette (FLAGSHIP CLUBHOUSE)

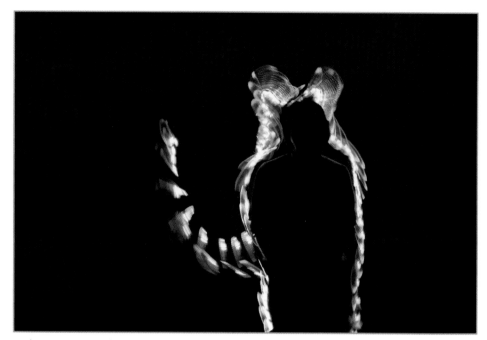

Light painting made with a multicolored light source (FLAGSHIP CLUBHOUSE)

GENERATING IDEAS

Ask makers to brainstorm ideas for light paintings that they want to create and techniques that they would like to try.

Brainstorming ideas for light paintings (FLAGSHIP CLUBHOUSE)

✂️⚙️ OPEN MAKE

During the Open Make time, encourage makers to experiment with new techniques.

For example, they can make wands with multiple lights. They can use blinking LEDs to make a dotted line. They can make thicker lines using materials that diffuse the light (such as clear straws, glue sticks, ping pong balls, and cotton balls).

Light painting using two different colors
(PUERTA 18 CLUBHOUSE, BUENOS AIRES, ARGENTINA)

Or they can paint the silhouette of an object by moving the light behind it.

Using a ping-pong ball to make a wand with diffused light (FLAGSHIP CLUBHOUSE)

Here are some additional ideas for light paintings:

- ▸ Draw sketches, words, or symbols.

- ▸ Trace objects or add to items already in the frame, such as by drawing hats or wings on a person who is standing still.

- ▸ Make group projects where each person draws a part of the drawing.

- ▸ Create an animation by taking a series of photos.

A small group light painting together (FLAGSHIP CLUBHOUSE)

Light painting created by tracing around a person (FLAGSHIP CLUBHOUSE)

LED lights in a jar (FLAGSHIP CLUBHOUSE)

Painting images using multiple light wands (PUERTA 18 CLUBHOUSE)

Drawing a word in the air (CLUBHOUSE ANNUAL CONFERENCE, DENVER, CO)

The Awa City Clubhouse in Whanganui, New Zealand, decided to arrange a light painting field trip. Clubhouse Coordinator Masina Kenworthy said,

We waited until it was dark outside and headed down to the river. We have a mighty river that flows through our city. At nighttime the lights from the road and town make beautiful reflections on the water. Our makers wanted to experiment with those lights in their photos. The river is constantly flowing, and they were able to capture the moving reflections in the photos.

Photo of lights along the river (AWA CITY CLUBHOUSE, WHANGANUI, NEW ZEALAND)

SHARING AND REFLECTING

We recommend making time to gather everyone into one group so they can reflect together and make notes in their journals. You can ask the makers to share how they tinkered with their light wands, how they achieved different effects, and how they were influenced by each other's ideas.

 # Journal Prompts

You can ask the following:

Did anything surprise you as you were working on this?

Did anyone help you with your project? What did you learn from them?

What would you do if you had more time?

📷 DOCUMENTING AND DISPLAYING

You can help makers document their process while building light wands and taking photos.

Work with the makers to put together a slideshow of all the finished photos, and then watch it as a group. You can also print out the light paintings and display them on the wall.

Light painting collage (YOUTH CONNECTIONS CLUBHOUSE)

STORY FROM THE CLUBHOUSE NETWORK: GLOWING JUMP ROPES

Janette Nelson, at the Sorenson Unity Center Clubhouse in Salt Lake City, UT, extended the activity by introducing electroluminescent (EL) wire.

Two of the girls participating in the activity jumped rope for ten seconds, creating ghostly effects with blue and green patterns. One of the girls had a red light in her hand, which added a splash of red.

They discovered that light painting was fun and easy to do. It was a good opportunity for all the participants to have creative freedom. Each maker was able to explore their own ideas and then also work as a team to make new patterns and mix colors.

One of the makers said, "I did not know that we could make LED light into art!"

Another said, "The thing that was hard was when the jump rope got stuck on my shoe. The thing that was fun was when my friend joined me. She saw me jump roping and wanted to do it too. . . . But I wish that we could do this again because it was fun—you guys should try it!"

Light painting created by jumping rope (SORENSON UNITY CENTER CLUBHOUSE, SALT LAKE, UT)

ADDITIONAL RESOURCES

▶ Light Painting guide from the Exploratorium's Tinkering Studio: tinkering.exploratorium.edu/light-painting

▶ A light painting overview from the Children's Museum of Houston: cnx.org/contents/_sOeT9AP@1/Light-Painting

Ideas for light painting projects...

Give It Form:

3D FORMS

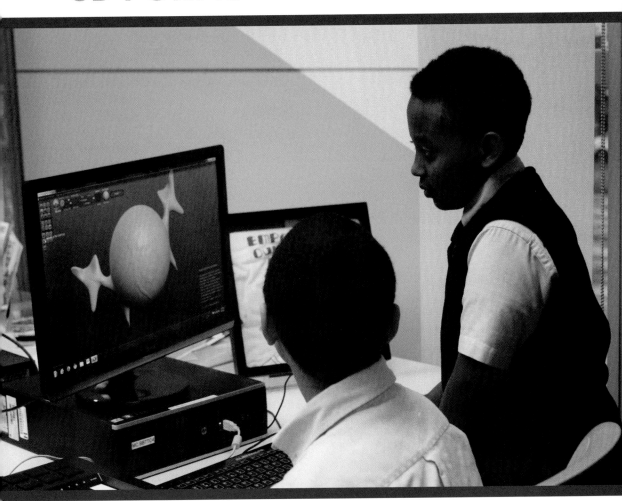

In this session, makers will design and build three-dimensional (3D) sculptures out of paper or clay. In the process, they will learn about transforming two-dimensional (2D) shapes into 3D forms. As an extension, you can introduce 3D modeling using Pixologic's Sculptris digital sculpting software.

SESSION GOALS

In this session, makers will

- Collaboratively design and create 3D sculptures using images as visual prompts.

- Engage in digital sculpting using Sculptris and design a model for 3D printing.

- Learn about transforming two-dimensional shapes into three-dimensional forms.

✔ GETTING READY

In this session you can introduce 3D design in two different ways: by paper modeling and by digital sculpting.

Make your own 3D projects as examples to help spark interest.

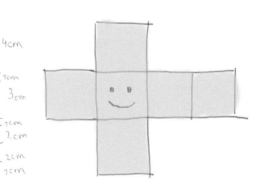

Design for a 3D paper toy (MUSEO DE LOS NIÑOS, BOGOTÁ, COLOMBIA)

MATERIALS

- ▶ Printouts of templates to make 3D forms:

    - ▶ Cube:
      bit.ly/start-making-cube

    - ▶ Pyramid:
      bit.ly/start-making-pyramid

    - ▶ Cylinder:
      bit.ly/start-making-cylinder

- ▶ Paper (plain, cardstock, grid, or graph)

- ▶ Modeling clay, such as Play-Doh

- ▶ Glue, clear tape or double-sided tape

- ▶ Rulers

- ▶ Crayons, pencils, and markers

- ▶ Sculptris 3D modeling software; download from
  pixologic.com/sculptris/

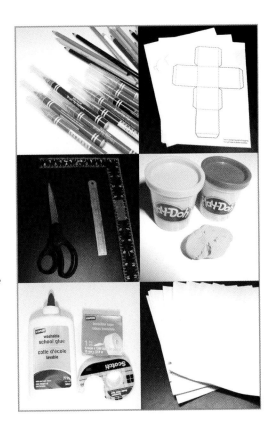

> ## SCULPTRIS
>
> Sculptris is a free digital sculpting interface from Pixologic that allows you to get started with 3D modeling. You can also export your sculpted model from Sculptris and print out your design using a 3D printer. Sculptris is based on the concept of modeling clay. Just as with modeling clay, you can pull, push, pinch, and twist a sphere of virtual clay and shape it into your desired form.

SPACE AND TOOLS SETUP

Lay out the basic materials for the session. You can also include other drawing tools and grid paper.

Install Sculptris software on each computer. If you plan to introduce digital sculpting, you'll need Sculptris software installed on each computer. If possible, arrange one computer for every two to three makers.

OPENING

We recommend beginning the session with two introductory activities to help makers start thinking about 2D shapes and 3D forms.

1. **Draw your own creature.**

   - ▶ Distribute scratch paper and ask makers to take five minutes to draw their own monster, animal, or cartoon character.

   - ▶ Prompt makers to give the creature a name and to list three adjectives to describe it.

   - ▶ Have makers introduce themselves and their creature to the group.

Character sketches (FLAGSHIP CLUBHOUSE, MUSEUM OF SCIENCE, BOSTON, MA)

2. Use 2D nets to make 3D forms.

In this starter activity, makers take a flat pattern (called a *2D net*) and make a pyramid, a cube, or a cylinder.

1. Ask the makers to get into groups of three. Give each group a template for a pyramid, a cube, and a cylinder (see Materials list).

2. Allow time for makers to cut out and fold each 2D shape into a 3D form.

Taping together a pyramid

3. Tape each of the 3D forms so that they hold together.

4. Ask the makers to think about the three forms. For example, "What does the pyramid remind you of?" "Where have you seen pyramids?" "Can you spot an object that is a pyramid in this room?"

2D NET

A *2D net* is a two-dimensional pattern that can be folded to create a three-dimensional form.

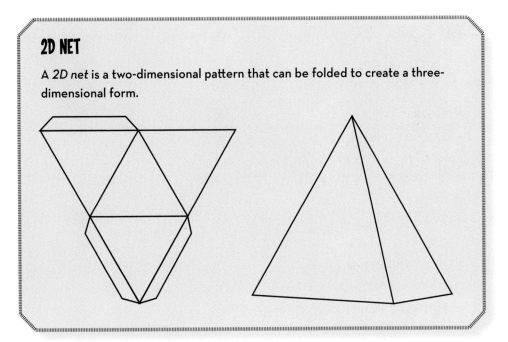

5. Then give the groups another short period of time (five minutes) to combine the three forms into a collaborative form, based on fun prompts such as "build the friendliest robot" or "form a spaceship."

6. Finally, give makers another short period of time to try to flatten the form back into a 2D net.

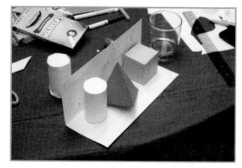

Cylinder, pyramid, and cube models made from their 2D net templates

⚡ DEMOING

Share an example that shows how a 2D drawing can lead to a 3D sculpture.

Sharing examples (FLAGSHIP CLUBHOUSE)

For example, you can show

- ▸ A drawing of a fantasy creature or other character

- ▸ A clay model of the character

- ▸ A digital 3D model of the character

- ▸ A 3D printed version of the character (optional)

3D model based on 2D sketch

Making a creature from clay (FLAGSHIP CLUBHOUSE)

BUILDING A FIRST PROJECT

For the first project, you can make either a paper sculpture or a clay sculpture.

Printed images of animals

PAPER SCULPTURES

Begin by having each maker choose what type of animal or character they would like to make. You can provide pictures of animals or creatures (such as a cat face or a fish). Give the makers the choice of using one of those characters or one of their own drawings.

Encourage them to identify and break down their image into basic shapes. For example, a cube or an octagon can be used for a head, a cylinder can be used for an arm, and pyramids can be used for ears or noses.

Ask the makers to design a 2D net based on their image.

Marking shapes on the image

Designing a 2D net to make a 3D form

The makers can later cut out, fold, and tape the 2D net to make a 3D form.

Making a 3D form of a cat face

3D form of a cat face made from an octagon and two pyramids

CLAY SCULPTURES

If you are planning to engage the makers in digital sculpting, start by using modeling clay.

1. In pairs, have the makers decide on a character that they would like to design out of clay. For example, they can make a hybrid animal, such as an "owl-bear," a "zebra-duck," or a "shark-elephant."

2. Pass out the modeling clay to the makers and encourage them to start sculpting their character.

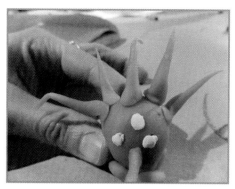

A character sculpted from clay (FLAGSHIP CLUBHOUSE)

Sculpting an animal from clay (FLAGSHIP CLUBHOUSE)

3. Encourage them to experiment with sculpting tools using clay.

4. Ask the makers to identify basic forms in their clay model, such as cones, cylinders, pyramids, spheres, and cubes.

Beginning to sculpt (FLAGSHIP CLUBHOUSE)

TAKING IT FURTHER

Now that the makers have experience designing 3D forms using physical materials, they are ready to explore the design of 3D models on computers.

DIGITAL SCULPTING

An inviting way to get started with 3D modeling is through digital sculpting—that is, forming a 3D shape by modeling it like clay.

Introduce Sculptris Software

Have all the makers gather near a computer screen and introduce Sculptris software. First, show an example project. Then, start a new, blank Sculptris file. Show how you can shape the virtual clay using different tools. Point out the key features (primary brush tools like "pull" and "pinch"). Pick up a clay form and show how surfaces change when you pinch and pull the clay.

Make Makers can look at their paper or clay models for inspiration. Ask them to think about the key shapes that make up the parts of their characters. They then can start making figures using the Sculptris tools to draw or shape features.

View After makers have had some time to experiment, show them how to zoom in and toggle the wireframe on and off; point out all the triangles that fit together to make the features they just pulled and pinched. It's geometry at work!

Introducing Sculptris software (FLAGSHIP CLUBHOUSE)

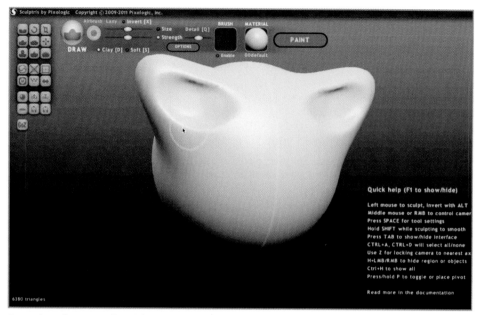

Digital sculpting with Sculptris (FLAGSHIP CLUBHOUSE)

Enhance Show makers how to manipulate the forms using other Sculptris tools, including selecting, moving, rotating, and scaling.

3D Print If you have access to a 3D printer, you can walk makers through the steps to convert their 3D models into printed form. If you do not have access to 3D printers, 3D printing services, such as Shapeways (shapeways.com), allow you to upload 3D files and get them printed.

Sculptris imports and exports .OBJ files. You'll need to convert to other formats, such as .STL files, to output to a 3D printer. To convert your Sculptris file, use a plug-in or conversion software, such as Meshmixer (meshmixer.com), to make it compatible with the make and model of the 3D printer. Make sure everything is connected and the model has no holes in it. You can make multiple prints of your design in different sizes and using different printing materials.

3D prints from Sculptris designs (FLAGSHIP CLUBHOUSE)

GENERATING IDEAS

Now that the makers have a basic understanding of how to create 3D forms, they can brainstorm other projects they can create.

To brainstorm ideas, ask makers the following questions:

▶ What are some things that you would like to change or add to the 3D sculpture you made?

▶ What is your next step? Sketch three ideas that you want to make using this process.

▶ Can you make your character friendly? Scary? How does the shape of the features communicate personality traits like this?

 OPEN MAKE

Here are some ideas to explore for 3D design during Open Make time:

▶ Design a character or avatar for a game.

▶ Create a self-portrait.

▶ Design your own origami or paper toys.

Sketching an avatar for Minecraft
(THUNDERBIRDS BRANCH BOYS & GIRLS CLUB, GUADALUPE, AZ)

Designing a self-portrait to build into a 3D sculpture (PUERTA 18 CLUBHOUSE, BUENOS AIRES, ARGENTINA)

Cutting out designs for 3D sculptures (PUERTA 18 CLUBHOUSE)

STORY FROM THE CLUBHOUSE NETWORK: SCULPTING JACK-O'-LANTERNS

Inspired by the approach of Halloween, the Gold Crown Clubhouse in Denver, CO, decided to use 3D-modeling tools to design jack-o'-lanterns. Clubhouse facilitator Victor Escobedo created a simple pumpkin template in Sculptris. The young people in the Clubhouse used the Sculptris digital sculpting tools to remix the file and make their own jack-o'-lantern designs. They then were able to 3D-print their projects. They were so thrilled by the results that they were motivated to make more 3D designs based on other holidays and personal interests.

A 3D model of a jack-o'-lantern designed in Sculptris (GOLD CROWN CLUBHOUSE, DENVER, CO)

SHARING AND REFLECTING

Encourage makers to take time to reflect on the process of creating 3D projects from 2D paper.

 Journal Prompts

You can ask the following:

What process did you use to make your character?

What challenges did you have going from 2D to 3D?

What else could you build with these tools and techniques?

DOCUMENTING AND DISPLAYING

Make a slideshow of the 2D shapes and 3D models, including photos or screenshots of the creations.

Create a stage for exhibiting the projects. You can also encourage makers to light up their 3D creations.

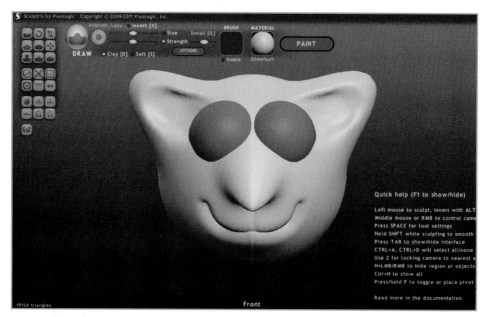

Screenshot of a 3D model in progress made using Sculptris

Lighting a 3D cube (YOUTH CONNECTIONS
CLUBHOUSE, LISMORE, AUSTRALIA)

ADDITIONAL RESOURCES

▶ Scultpris features:
pixologic.com/sculptris/features/

▶ Shapeways 3D printing services:
shapeways.com

▶ Designing paper toys:
bit.ly/papertoydesign

Sketches of my 3D creations

Change the Move:
ART-MAKING BOTS

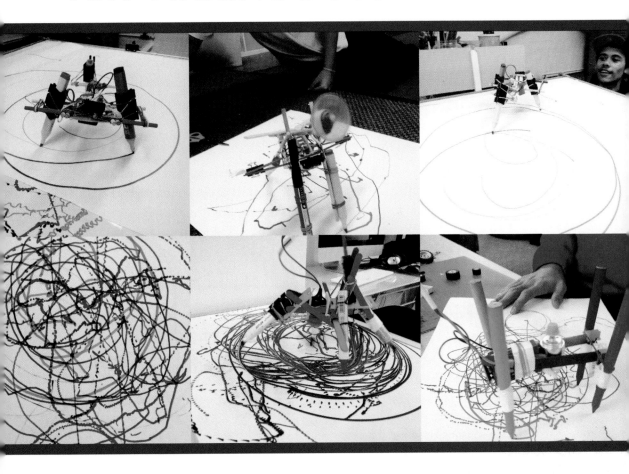

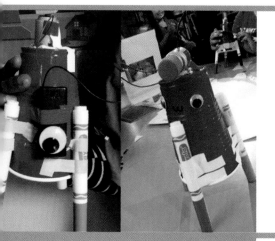

Makers will build an art-making "bot" that draws as it moves. They'll use parts from an electric toothbrush or other small device. Once they have built their bot, they can observe and then change its movements to draw interesting patterns on paper.

SESSION GOALS

In this session, makers will

▶ Deconstruct a small household device to see how it works and then repurpose its parts.

▶ Build a vibrating art bot that makes drawings.

▶ Investigate physical properties of motion, exploring how attaching different weights to a motor shaft changes the way a bot moves and draws.

✔ GETTING READY

For this session, makers will take apart an electric toothbrush or another small motorized device and give it a new purpose—reusing the electrical components to make an art-making bot.

To prepare for the session, it is important to try building a bot first. By trying it yourself, you can make sure that the materials you choose will work well for the activity.

IS IT A ROBOT?

Although your art-making bot will have a motor and move, it isn't technically a real robot, because it isn't controlled by a computer program and doesn't have sensors. However, it may keep moving on its own—even when it hits a wall!

MATERIALS

- ▸ A small toy or household device that contains a circuit with a motor, battery power, and a switch (such as a disposable electric toothbrush, a mini-handheld electric fan, or a toy car)

- ▸ A plastic cup or bowl

- ▸ Drawing tools, such as markers, crayons, or chalk (avoid permanent markers)

- ▸ Fasteners (such as masking or electrical tape, double-sided foam tape, pipe cleaners, or rubber bands)

- ▸ Cork or old eraser

- ▸ Art supplies such as googly eyes or pipe cleaners

- ▸ Wires or copper tape and a wire cutter/stripper

- ▸ Butcher paper or large sheets of art paper

- ▸ Tools for deconstructing machines safely, such as pliers, screwdrivers, scissors, and safety goggles

★ FACILITATION TIP

Work with Locally Available Materials

In some areas, it may not be easy to find inexpensive electric toothbrushes or hand-held fans. You can look in local shops for other small items to decon-struct, or purchase small motors and battery holders from an online elec-tronics store.

Masina Kenworthy from the Awa City Clubhouse in Whanganui, New Zealand, described her experience looking for materials:

> We don't have inexpensive toothbrushes in New Zealand, so we used a little hand fan, because they came with on/off switches. The members had to be really creative to come up with ways to get the bot to move, as the motor didn't vibrate or have as much power as the toothbrushes would.

Small handheld fan

SPACE AND TOOLS SETUP

You will need a lot of room on work tables for building the art-making bots. You will also need a wide space on a table or the floor to test your art creations. Keep in mind that your bots' drawings will be more interesting if there are obstacles or walls that they can bump into, causing them to change direction. Cover the space with a few layers of butcher or art paper. If you are going to use markers, make sure the ink doesn't bleed through the paper onto your floor or table.

Building art-making bots (TECNOCENTRO SOMOS PACÍFICO CLUBHOUSE, CALI, COLOMBIA)

OPENING

For this session, makers will need to lean on each other for brainstorming, collaboration, experimentation, and remixing. You could set the tone for teamwork by opening this session with a classic Clubhouse group brainstorm known as a LEGO Doodle. The process of LEGO Doodling encourages makers to tinker, remix, experiment together, and dive in with a spirit of playful inquiry and collaboration.

LEGO DOODLE

Introduce the activity by asking whether any of the makers ever doodle on paper. Then, explain that in this activity, everyone will try doodling using LEGO.

1. Invite everyone to sit in a circle around a table, and dump a big, messy pile of LEGO blocks (or other materials) in the center, so that everyone can reach the pile.

2. Ask the makers to pick up a few LEGO pieces that attract their attention. Tell them not to worry about what they are building. They can just fit the pieces together however they would like.

3. After a timed period, such as one minute, prompt the makers to pass their LEGO Doodle creation to the person to their right in the circle. That person then will add several more pieces to the object.

4. Repeat the process all the way around the circle. Return each creation back to the original Doodler.

5. Invite everyone to look at their object to see if a title comes to mind. Ask them to share their title with the group.

6. Reflect on how the Doodle experience made them feel as creators, which aspects made them feel motivated, and which made them feel challenged or "stuck." How did working at a fast pace, receiving a Doodle from their neighbor, or seeing their original creation evolve help or interfere with their creative process?

LEGO Doodle activity (HENNEPIN COUNTY LIBRARY BEST BUY TEEN TECH CENTER, MINNEAPOLIS, MN)

⚡ DEMOING

Make a few demo art-making bots of your own first, and have some examples ready to show. It's good to show what the electronic device looked like originally, so your makers can compare before and after versions.

You can also show the makers different kinds of drawings created by your bots. Ask questions, such as these:

▶ What parts can you identify on the art-making bot that came from the original device?

▶ Which of these bots do you think made the most interesting pattern?

Examples of bots making drawings (YOUTH CONNECTIONS CLUBHOUSE, LISMORE, AUSTRALIA)

⇨ BUILDING A FIRST PROJECT

This activity is inspired by the Scribbling Machines activity from the Exploratorium's Tinkering Studio and the Art Bot activity on the Science Buddies website (see Additional Resources).

Open the device and find the circuit. Ask the makers to pair up or get in groups, and give each group a device and a set of tools for taking it apart (such as scissors, wire cutters, and pliers). Also provide them with safety goggles to use when taking the device apart.

Encourage them to take the devices apart carefully so that they avoid damaging the parts, because they will need to use them to build the bots.

Once the casings (usually plastic) are off, prompt makers to document in photos or sketches how the device is wired. They can lay out the parts on the table. You may want to suggest that they label the parts on the paper underneath, so they can remember the names of the parts and what they do.

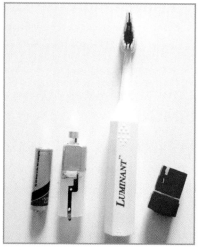

A disposable electric toothbrush, deconstructed

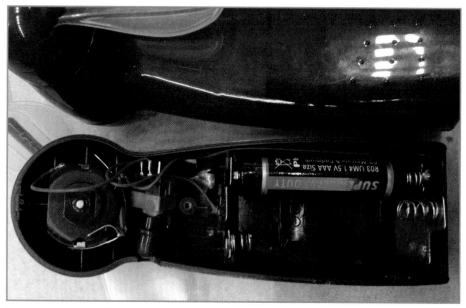

Taking apart a small hand fan with a motor and battery

Reconnect the circuit. Ask the makers to try to reconnect the circuit and make the motor spin once it is out of the casing.

Reconnecting the circuit inside a disposable electric toothbrush using copper tape

Now makers are ready to design and build their art bots. You can help guide them through the following steps.

1. **Gather materials.**

   Gather and lay out the basic parts for an art bot: a body (such as a plastic cup), legs (markers or other drawing tools), and joints (fasteners).

2. **Add the motor and battery.**

   If you're using a cup as the body for your bot, turn it over so the flat part is on top. Then fasten the circuit with the motor and a battery to the top.

   Tape or attach the power source (usually a battery or battery pack) to the bot's body somewhere near the motor. Make sure connections are tight and parts are stable on the base. If the battery doesn't have a holder, wrap a rubber band around the ends to keep the wires in place.

Motor powered with a battery placed on the base of a plastic cup

3. Add a wobble.

If you want your bot to move around, you'll need to throw its movements off center. Adding a weight to the rotating motor makes it vibrate, which will help it wobble or wiggle, and thus move in interesting patterns. The motor inside a disposable electric toothbrush already has some weight on the shaft that makes it vibrate. If

Motor with cork attached to its shaft

your motor does not have additional weight, you can press a cork, eraser, popsicle stick, or glue stick onto the rotating shaft of the motor. You can also add leftover craft supplies (such as beads) onto your weight—an off-center weight swinging from the shaft will make the bot shake and hop even more.

4. Give your bot some legs to stand on.

To add some legs, tape at least three markers around the rim of the cup, pointing down. You can also use crayons, chalk, pens, or dry erase markers.

Motor powered with a battery placed on the base of a plastic cup

---

 FACILITATION TIP

Make Observations, Then Tinker

Encourage the makers to take notes on how things are working (or not!) and to try something new. Model the act of tinkering and the process of scientific inquiry: after everyone creates their first bot, invite them to observe how it moves and ask questions about why. Then, suggest that they change the size or location of the weight and notice how this affects the way the bot moves or wobbles and how it draws.

5. Add an on/off switch.

Check the wire connections of your battery to your motor. If you just tape these wires permanently to the battery, you won't be able to turn off the motion. Stop and think about how to make a simple or triggered switch using parts left over from the original device.

Using a metal paper clip as an on and off switch

6. Create a persona for the bot.

Add some personality to your art-making bot by adding eyes, pipe cleaners, and bells. Set your creation loose to move around.

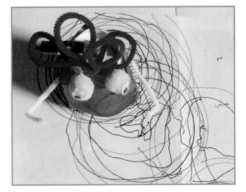

A complete art-making bot

TROUBLESHOOTING TIPS

Here are some tips shared by Start Making! facilitators at various Clubhouses:

"Too much glue causes motor problems sometimes—it's a small fix to remove excess dry glue and try again."

ANDY DERILUS, YWCA Clubhouse, Miami, FL

"Get the proper amount of counterweight to keep the bots from moving too fast."

SHANICE JOHNSON, Thunderbirds Branch Boys & Girls Club, Guadalupe, AZ

"Rewiring the battery to the motor doesn't always work on the first try, especially when we realized the wire was too thick so it was making it hard to get electrical current to the battery. We just had to go purchase thinner wire and it made it much easier."

ADRIANA FLOWERS, La Alameda Plaza Clubhouse, Walnut Park, CA

 ## TAKING IT FURTHER

Once the makers have successfully created their art-making bots, encourage them to bring all the bots to a common place, such as the community table or a bot "ring" on the floor, and get them to start drawing together. Step back and encourage the makers to observe the movements of the bots.

As everyone watches their bots together, ask questions such as these:

▶ How are the designs of the bots different? For example, do they differ in the placement of the motor, the materials used, and how parts are attached?

▶ How does the design of the bot affect the patterns it draws?

▶ What might you change to draw different kinds of patterns?

GENERATING IDEAS

Explore how individual bots can interact to make collaborative designs. Prompt makers to create interactions between their bots to make more complex and interesting drawings together.

 # OPEN MAKE

During Open Make time, encourage makers to keep tinkering and refining their bots.

>  ## ★ FACILITATION TIP
> ### Encourage Problem Solving
> You can support makers in troubleshooting their art bots by sitting down with them and asking questions.
>
> ▶ If the bot keeps falling over, ask: What happens if you spread the markers out further?
>
> ▶ If the bot keeps going in only one direction, ask: What happens if you change the direction the motor is facing?
>
> ▶ If the bot keeps falling apart, ask: How can you make it more robust?
>
>
> Experimental design for an art bot

Here are a few suggestions you could make:

▶ Attach another motor and observe the new drawing patterns.

▶ Add more markers or other drawing instruments.

▶ Add a sensor to turn the motor on (such as a photoresistor or reed switch).

▶ Add other design features (such as making the bot look like a fictional character).

An art bot with six markers for legs (AWA CITY CLUBHOUSE, WHANGANUI, NEW ZEALAND)

SHARING AND REFLECTING

Gather for a reflection circle. You can prompt makers to document what they learned in their journals.

✎ Journal Prompts

You can ask the following:

What did you like best about this activity?

What is something that you noticed or learned?

How might you apply what you learned somewhere else?

📷 DOCUMENTING AND DISPLAYING

Encourage makers to stop and document the steps along the way as they build each version of their bots. Invite them to take digital photos or to make sketches in their notebooks or journals.

Showcasing an art-making bot in action

Ask makers to take turns building the bot and capturing their progress with photos. Invite them to take notes about which changes worked or failed, and about which materials worked best. Later, they can use these in-process photos to tell a story of the "birth" of their art-making bot.

Makers can look at the artwork that the bots created and choose their favorite drawings to display.

STORY FROM THE COMPUTER NETWORK: ART BOT FASHION SHOW

Members from the Sacramento Food Bank & Family Services Clubhouse in Sacramento, California, drew on their love of fashion to put together an "art bot" showcase with a fashion runway theme.

Clubhouse Coordinator Kelly Ann Adams and the other facilitators discovered that many of their makers were especially interested in fashion. So, during the open make time, they encouraged the makers to design a costume for their bots. The makers then tinkered with the bot movement so the bots could "walk" down a ramp covered in paper. They chose their favorite art-bot drawings and put them on display.

At their final showcase, the makers hosted a fashion show, where each art-making bot walked down the runway drawing unique patterns. The makers then showed their family and friends how their bots worked. The makers auctioned off the art-bot drawings as a fundraiser for future Start Making! activities.

ADDITIONAL RESOURCES

▶ Exploratorium's Scribbling Machines:
   tinkering.exploratorium.edu/scribbling-machines

▶ Science Buddies—Art Bot:
   bit.ly/science-buddies-art-bot

▶ Switch Basics- Sparkfun:
   learn.sparkfun.com/tutorials/switch-basics

What's next for my art-making bot?

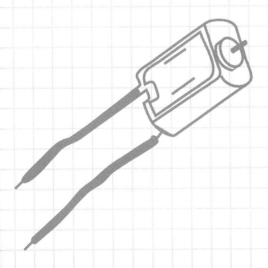

Sew the Circuits:

E-TEXTILES

Makers will create a bracelet or bookmark with programmed light effects. In the process, they will learn how to sew circuits with sewable LED lights, conductive thread, and the LilyTiny microcontroller, which can turn the LED lights on and off in surprising and interesting ways.

SESSION GOALS

In this session, makers will

▶ Make an electronic circuit with interesting light effects using sewable LED lights and a preprogrammed board called a LilyTiny.

▶ Design a light-up project such as a bracelet or a bookmark.

▶ Explore what they can create by combining sewing and electronics, as an introduction to a new area known as soft circuits or e-textiles.

✔ GETTING READY

You and your makers are about to explore e-textiles, which can inspire makers to learn both sewing and electronic skills. The projects make use of new materials, such as conductive thread and the LilyTiny board, which is designed for sewing patterns of lights onto fabrics.

To prepare for the session, gather the materials. Then, try making a bracelet and a bookmark yourself to share as examples.

Display of craft and electronic materials put together by facilitators before the session

MATERIALS

- ▶ Felt
- ▶ Needles
- ▶ Conductive thread
- ▶ Regular thread
- ▶ LilyPad sewable LEDs
- ▶ LilyTiny
- ▶ LilyPad sewable battery holder
- ▶ Coin batteries, 3V (such as the CR2032)
- ▶ Alligator clips (optional)

To find out how to order LilyPad parts, see lilypadarduino.org or order directly from sparkfun.com/products/10899.

LILYTINY

The LilyTiny is a tiny LilyPad board that can make LEDs twinkle and blink in fun patterns. It is a sewable, preprogrammed board that makes LEDs light up in fading, twinkling, blinking, and flickering patterns.

A LilyTiny board has four pins to which you can attach LEDs. Each pin has a different pattern.

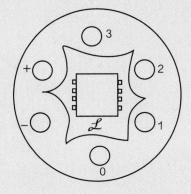

► Pin O fades in and out.

► Pin 1 thumps in a heartbeat pattern.

► Pin 2 steadily blinks on and off.

► ThePin 3 flickers like an artificial candle.

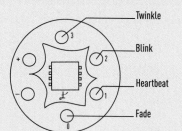

To use any of these patterns, sew the positive (+) side of the LED board to one of these LilyTiny pins and the negative (–) side of the LED board to the LilyTiny's negative (–) pin.

— from the LilyPad site: lilypadarduino.org/?p=523

SPACE AND TOOLS SETUP

You can organize and display all the materials, putting together the familiar craft supplies alongside the new electronic materials.

Be ready to show makers how to thread a needle and sew a backstitch. You can set up sewing demos or sewing stations to help makers who don't have previous experience sewing by hand.

OPENING

As makers arrive, you can invite them to walk around and touch the different objects that they are going to explore in the session. Ask them to make mental notes of what is surprising or interesting to them.

If your group does not have much previous experience with sewing, start with a short sewing activity or "sewing circle." Show how to do a basic stitch using regular thread before you introduce the conductive thread. Conductive thread is expensive and can be a bit difficult to work with for first-timers.

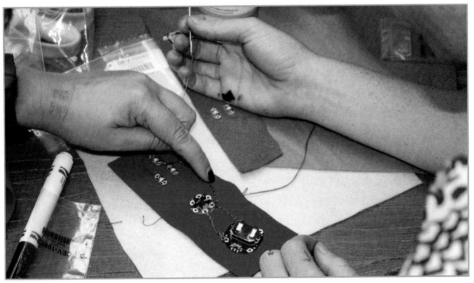

Start Making! facilitator demonstrating how to sew (CEDES CLUBHOUSE IN ALAJUELITA, SAN JOSÉ, COSTA RICA)

For sewing electrical connections with conductive thread, the back-stitch works the best. This stitch takes time, but it makes a solid connection. You can find many tutorials for how to do a backstitch online or in any introductory sewing book. Check the "Additional Resources" section later in this chapter for a recommended backstitch video tutorial.

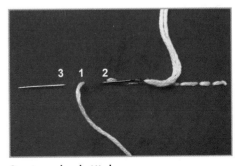

Sewing a backstitch

⚡ DEMOING

Show an example of a LilyTiny project that you designed. Share your design process with the makers and highlight the parts they will be using for their bracelets or bookmarks.

LilyTiny boards are already programmed. Introduce the different patterns of the LED lights attached to the LilyTiny boards. Makers will remix these light patterns in their own projects.

Facilitators explaining how to sew circuits
(COMPADRE BOYS & GIRLS CLUB CLUBHOUSE, CHANDLER, AZ)

Soft circuits project example

★ FACILITATION TIP

Encourage Material Exploration

Making a soft circuit is just like creating a paper circuit, only with conductive thread instead of copper tape. Encourage your makers to play with the conductive thread. They can try making a paper circuit by taping conductive thread instead of copper tape. They can also compare two circuits, one with copper tape and one with conductive thread, to see which one makes the LED brighter. Another way to explore this activity is for makers to connect two pieces of conductive thread by knotting them. Do they still transmit electricity? What happens to the light? Makers can use small leftover pieces of conductive thread for these explorations.

The thread and fabric can be hard to work with, since the thread gets tangled easily and the fabric doesn't always lie flat. Also, conductive thread has a tendency to fray, so cut generous amounts of thread.

⇨ BUILDING A FIRST PROJECT

You can use the following steps for making the soft circuits to help guide the makers. This activity is inspired by a project described in more detail in the book *Sew Electric* by Leah Buechley, Kanjun Qiu, and Sonja de Boer (2013). You can find more information at sewelectric.org.

Sewing electronic bookmarks and bracelets (SWICN CLUBHOUSE, DUBLIN, IRELAND)

1. **Cut the fabric.**

   Cut rectangles of felt, roughly 3×8 inches (or about 8×20 centimeters). If you want to make a bracelet, trim the length down to your wrist size with about an inch of overlap. Use snaps to create a bracelet. Otherwise, you can leave your rectangle flat if you want to make a bookmark.

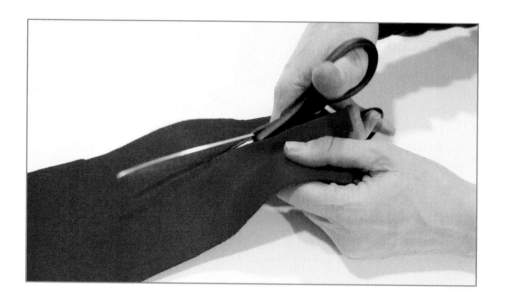

2. Sketch a project design.

Make a sketch of your design for the bracelet/bookmark. You can decide how many LEDs you want to use for your design and mark the positions of the LEDs in your drawing.

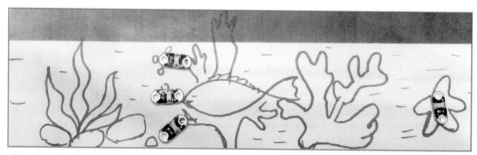

Sketch of a light-up bracelet with LEDs placed on top to plan which parts will light up

3. Choose the light pattern.

As mentioned earlier, LilyTiny has four pins with different patterns programmed into each one. Pin 0 makes an LED fade in and out, pin 1 makes the LED light up in a heartbeat pattern, and pins 2 and 3 make the light steadily blink and twinkle, respectively.

Choose the LED pattern that you want to add to your bracelet/bookmark. You can have all the LEDs light up in the same style by connecting them all to one pin, or you can attach each light to a different pin for multiple light patterns.

4. **Mark the position of the battery holder and LilyTiny in your design drawing.**

   In your drawing, mark the positions of the battery holder and the LilyTiny microcontroller. Sketch the connections between the battery negative (–) terminal and the LilyTiny negative (–) terminal. You can use different color pens to distinguish the connections to the positive and negative terminals.

5. **Draw the circuit connections on top of the project design drawing.**

   Now, draw the connections between the positive (+) sides of

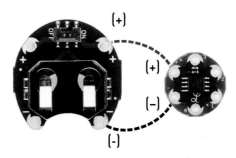

Circuit layout showing connections between battery holder and LilyTiny

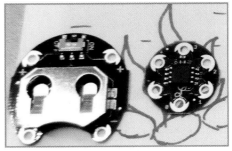

Sketch of a light-up project, with the battery holder and LilyTiny placed on top of the drawing

your LEDs and the LilyTiny pins. The following diagram shows connections from pins 1 and 2 to the LEDs.

Connect the negative (–) sides of both the LEDs to the negative (–) pin of the LilyTiny board or the negative (–) side of the battery, as shown in the figure.

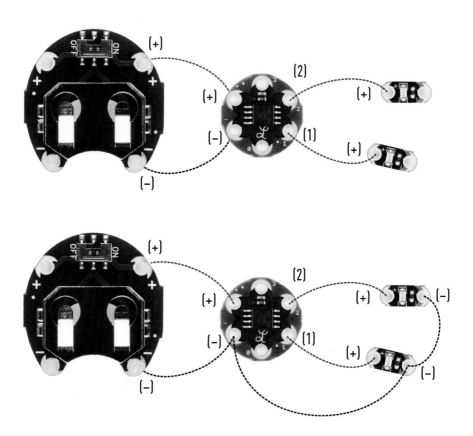

Make Observations, Then Tinker

If you want all the LEDs to have the same effect, connect all the positive terminals of the LEDs to the pin that you chose and all the negative terminals of the LED to the negative terminal of the battery.

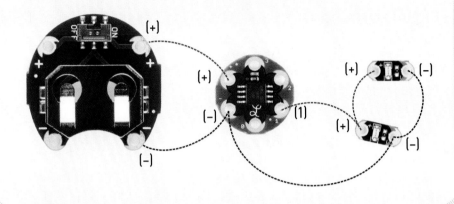

6. Sew the circuit.

Once you have marked the electrical connections on the bracelet/bookmark design, cut out your smaller elements. Then place the components on the fabric and start sewing.

To make it a little easier, you can glue the battery holder, LilyTiny board, and LEDs to the fabric before sewing with the conductive thread. Just make sure that there is no glue on the metal coating.

Stitch your circuit with the conductive thread. Attach the battery to the LilyTiny first, and then make the connections to the LEDs.

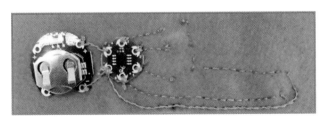

Sewing the circuit

7. Test the circuit.

Add a 3V battery. Then turn on the switch on the LilyTiny board.

Circuit sewn on felt

8. Embellish your project.

Refer to your initial drawing and complete your bookmark or bracelet design with craft materials. You might also want to try some appliqué embroidery. Felt fabric diffuses the light very well. You might also use pom-pom balls and googly eyes to decorate.

A blinking light-up bracelet

Sewing might not be the first technical skill you associate with a STEM or engineering activity, but there are many reasons why it should be! Part of the fun of Maker culture is that it embraces design and crafting as much as science and engineering skills. This is especially true of projects that incorporate low-tech materials such as thread, cloth, and sequins combined with high-tech wearable components such as conductive thread, programmable microcontrollers, and LEDs.

In the Start Making! program, we embrace both of these aspects, especially when they present an opportunity to engage youth and adults who may be reluctant to participate in a technology-based program. When facilitated well, sewn-circuit activities disrupt stereotypes based on gender, age, and expertise. The activity can motivate anyone interested in crafting, art, or fashion to become excited about materials science and coding.

Rodrigo Castellon sewing electrical circuits (GILA RIVER BOYS & GIRLS CLUB, SACATON, AZ)

Clubhouse Coordinator Rodrigo Castellon at the Gila River Boys & Girls Club in Sacaton, AZ, expressed his newfound interest in sewing by designing his own tie as a soft circuit project. The youth in his Start Making! program were inspired by his enthusiasm and ingenuity, and many also made light-up wearables as a result!

 ## TAKING IT FURTHER

You can encourage the makers to try using all the pins on the LilyTiny in the next iteration of their soft circuit projects.

GENERATING IDEAS

Encourage the makers to share their ideas for projects using LilyTiny boards.

To brainstorm ideas, ask makers the following questions:

- ▶ What would you change in your design if you were to make something like this again?

- ▶ Brainstorm ideas for designing more light behaviors or patterns.

- ▶ What would you do if you got a chance to change the pattern of lights?

A facilitator and a young maker discussing ideas while sewing together (FLAG-SHIP CLUBHOUSE, MUSEUM OF SCIENCE, BOSTON, MA)

 **OPEN MAKE**

Explore these ideas in the Open Make time.

Express something personal. Make an e-textile project that shares something special about you. It could present your hobbies, things that you like, or the place where you come from.

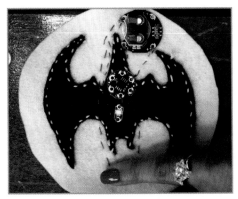 

E-textile projects made using LilyTiny boards, each project representing something special about the creator (CLUBHOUSE NETWORK TEEN SUMMIT, BOSTON, MA)

Remix this circuit. Brainstorm ways to create a collaborative project combining two or more LilyTiny boards.

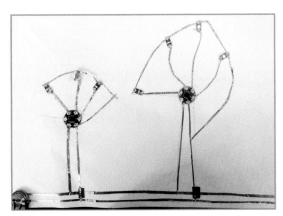

Project prototype with two LilyTiny boards powered by a single 3V battery

Make a 3D soft circuit. Make your own stuffed toy and add lights to your creation using conductive thread.

Stuffed toys with lights (JORDAN BOYS & GIRLS CLUB CLUBHOUSE, CHELSEA, MA [LEFT]; CLUBHOUSE ANNUAL CONFERENCE, DENVER, CO [RIGHT])

💬 SHARING AND REFLECTING

Share everyone's projects with the group. Encourage the makers to write their session reflections in their journals.

Journal Prompts

You can ask the following:

What problems did you need to solve?

What did you like best about this activity?

What do you want to learn more about?

📷 DOCUMENTING AND DISPLAYING

Encourage your makers to share the design process behind their soft circuit projects. Ask them to create a display board for their project with their bracelet/bookmark drawings and circuit sketches. If possible, include some photos of your makers wearing their bracelets or with their light-up bookmarks inside a book. Encourage them to make notes on their display board to capture their moments of joy and frustration while working on this project and the ideas they have for taking it further.

All the annotated display boards can be posted on a wall or table.

Makers can also use media tools like photo slideshows or videos.

Soft circuit bookmarks and bracelets (SORENSON UNITY CENTER CLUBHOUSE, SALT LAKE CITY, UT)

Makers at the Sorenson Unity Center Clubhouse in Salt Lake City, UT, documented their process of creating soft circuit projects, including photos. They also shared reflections about their experiences:

> "It was hard because I was running out of thread. It was fun because I learned and had new experiences. I would like to light up a car."

> "I liked sewing my project. It was fun. I would like to make a star that lights up."

> "I liked the project. I would like to make a haunted house."

> "I did not like it because I don't like sewing. I would like to do a car with a solar panel."

"It's beautiful art. I would like to make a video game controller."

"The part that I thought was hard was the conducting wire—making sure that it wasn't touching the other and things like that—but it was fun because later the work would be all cool and creative."

Shared by Janette Nelson, Clubhouse Coordinator, Sorensen Unity Center

ADDITIONAL RESOURCES

▶ Sew Electric:
sewelectric.org

▶ LilyTiny:
lilypadarduino.org/?p=523

▶ Mary Corbet's Needle 'n Thread channel about the backstitch:
http://bit.ly/backstitch-demo-tutorial

Three things I found the most challenging in this session are...

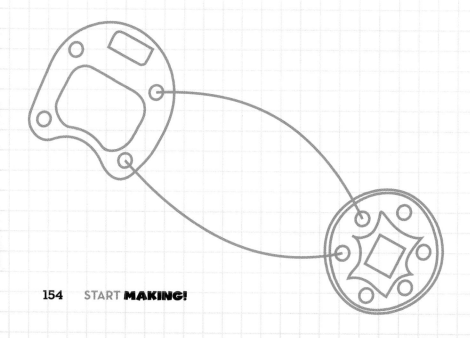

Final Open Make:
PERSONALIZED PROJECTS

Now that the makers have completed all the activity sessions, we recommend offering one or more Open Make sessions, dedicated to working on final projects. During these Open Make sessions, support makers as they imagine, design, and build personalized projects.

SESSION GOALS

In this session, makers will

▸ Plan and create a culminating personal project, either individually or in small groups.

▸ Build on learning and ideas from earlier sessions, applying STEM concepts and techniques in a new medium or context.

▸ Dive deeper into earlier interests or incorporate new interests, media, or tools.

✔ GETTING READY

Set up the space so that your makers will be able to brainstorm and then build personal projects. Arrange the materials and tools so they are accessible and organized to help makers easily find and utilize everything safely.

Lay out materials from the previous sessions in addition to the Open Make box. You may want to add additional materials and tools, such as those listed in the "Session Flow" chapter in the section called "Open Make Box."

 FACILITATION TIP

Creating a Respectful Environment

Before makers dive into Open Make work, you may want to remind them to show respect for the space, materials, shared tools, and each other. You can refer back to the "I Am a Maker" diagram from Part I and facilitate a group discussion on ways to support each other while brainstorming, making, and sharing projects.

Working on Open Make projects (YOUTH CONNECTIONS CLUBHOUSE, LISMORE, AUSTRALIA)

GENERATING IDEAS

Start by brainstorming what projects the makers would like to work on, either individually, in pairs, or in small groups. They may choose to build on a project idea they worked on in a previous session or start something new.

Idea wall (AWA CITY CLUBHOUSE, WHANGANUI, NEW ZEALAND)

Suggest that makers look at ideas they generated during previous sessions, which they documented in their design journals and on the idea wall. They can look through previous journal entries to consider the materials, tools, and projects they enjoyed making and that they might want to explore further.

Give them sticky notes to write down the ideas they would like to pursue for their final projects. In addition to brainstorming *what* they want to make, they can also write notes about *how* they will make it—by listing different tools and techniques and any questions they have about the process. They can also write notes about *why* they want to make something, for example, to solve a problem, for fun, or to help someone. Makers can help each other in this process, adding notes and suggestions.

Sample project ideas

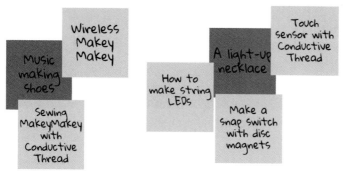

Project ideas with notes about process

 ## BUILDING OPEN MAKE PROJECTS

Remember that Open Make is designated time when makers create their own projects, building on starter activities and pursuing a personal interest, question, or design challenge. Help your makers progress from a basic project they each have done in the previous sessions to a more complex and advanced project.

Developing an Open Make project (GILA RIVER BOYS & GIRLS CLUB, SACATON, AZ)

You can suggest that they build on one of their previous projects or combine several types of projects (for example, making a musical project that lights up, or combining paper circuits with a MaKey MaKey and Scratch). Encourage makers to sketch out their project ideas and then take photos and videos to document their process as they work on their projects.

Experimenting with a motor (TECNO-CENTRO SOMOS PACÍFICO CLUBHOUSE, CALI, COLOMBIA)

⭐ **FACILITATION TIP**

Start with Imagination, Then Build Prototypes

Over the course of the Start Making! program at the Thunderbirds Branch Boys & Girls Club, in Guadalupe, AZ, everyone was able to create and take part in fun, interactive making activities. The Open Make session was the highlight of the program, as it allowed the youth to create and facilitate their own projects.

Clubhouse Coordinator Shanice Johnson reflected on their Open Make session:

The importance of the Open Make is to give youth a grounding in creativity—to allow makers to think for themselves and to tap into their ingenuity.

We started the facilitation by having each of the youth brainstorm their ideas. The initial question was, "What do you see yourself making? Imagine your creation."

Then the youth started utilizing any supplies to build and create their own works. I encouraged each of them to see and begin the frameworks for each of their projects, for example, drawing out an outline, or using pipe cleaners or big sheets of paper to begin and build prototypes.

SHARING AND REFLECTING

Although your makers might want to keep building, be sure to make time for everyone to pause and document their Open Make projects in their Start Making! journals.

You can also organize group reflections such as a group critique, a mini-showcase, or an internal Maker Faire as a dress rehearsal for a final Show and Share community showcase. Makers can share versions of their personal projects, along with their prototypes and the materials they used, and get feedback from each other.

Light box paper circuit project (GUM SPRINGS COMMUNITY CENTER CLUBHOUSE, ALEXANDRIA, VA)

Group critique (FLAGSHIP CLUBHOUSE, MUSEUM OF SCIENCE, BOSTON, MA)

Journal Prompts

You can ask the following:

What was your project idea?

What inspired you to pick this project?

What was your process for creating the project?

What was the hardest part?

What problems did you solve?

What is your favorite aspect of your project?

What are you most proud of?

What might you want to do in the future?

📷 DOCUMENTING AND DISPLAYING

Consider both low-tech and high-tech ways for makers to document and show off their work, including placing projects on shelves, bulletin boards, and poster boards, and documenting online.

Encourage makers to be persistent in building and remixing, to find new solutions, and to overcome roadblocks.

Give reminders so makers are aware of how much time they have to work on their projects. Reserve time for them to document their projects (as well as help clean up!). See Session 8 for ideas on how to plan and get ready to share their projects at a Community Showcase.

Documenting a project (HENNEPIN COUNTY LIBRARY BEST BUY TEEN TECH CENTER, MINNEAPOLIS, MN)

STORIES FROM THE CLUBHOUSE NETWORK: TWO OPEN MAKE EXPERIENCES

Open Make projects can lead in many different directions based on makers' interests. Here are two examples of projects created during Open Make sessions.

Alien Autopsy Game

When makers were invited to explore any activity or tool during Open Make time, one girl immediately chose MaKey MaKey. She started by playing around with Play-Doh, making tiny aliens. Then a sample project on the shelf caught her eye: a Valentine candy box version of the board game *Operation*. The goal of the game was to pick up a piece of candy from the candy box without touching the edge. The makers of that game had used a MaKey MaKey board and programmed a project in Scratch, so that it would buzz if the player touched the edge of the box.

Meanwhile, one of the staff members had brought in a large refrigerator-sized cardboard box. Inspired by the Valentine candy box, everyone brainstormed the idea of making a life-sized version of one of the tiny Play-Doh aliens. Mentors helped the young maker sketch the body of the alien on the box and think about how she could use the materials and tools provided to create her own game. Eventually she created a MaKey MaKey interface with keys at different parts of the alien's body, and she used Scratch to program her game. When touched, each body part emitted a different sound, like a pumping heart or a stomach growl.

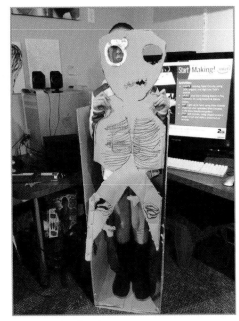

Sharing the final project (FLAGSHIP CLUBHOUSE)

Maze for a Pet

During the Open Make session at the Youth Connections Clubhouse in Lismore, Australia, one young maker wanted to design a maze for his pet rat, Beauty. Because he had a special bond with his rat, he was motivated to spend time and attention working out the details of the project. With support from mentors, he drew a maze on paper, and then built the floor, walls, and

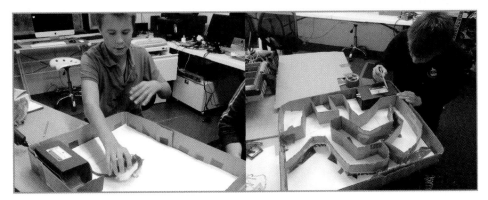

Building a maze (YOUTH CONNECTIONS CLUBHOUSE)

ceiling of the maze using corrugated cardboard. He added a box for Beauty to nest in. Then he wired up some colored LEDs to 3V batteries and made a lighted passage through the maze. He was excited to let his rat explore the maze and was proud of the project he had created.

REFLECTIONS FROM START MAKING! FACILITATORS

Here are a few reflections from Clubhouse Coordinators on facilitating Open Make sessions.

TRY IT AND SEE

To me, Open Make time is about igniting the curiosity for experimentation within our youth so that their interest in figuring out how is stronger than their aversion to failure or not knowing. Yes, we need to show them what can be done and how to use tools they've never seen, but I think our goal should be eventually to be unnecessary to their exploration. Makers don't need to know the way to do something; they just need the curiosity to figure out a way, so I think our job is to instill that mindset more than specific skills.

—Anna Fleming, Gold Crown Clubhouse, Denver, CO

Facilitator helping during an Open Make session (CASA DE LA JUVENTUD CLUBHOUSE, MORA, COSTA RICA)

CONNECTING THE DOTS

By engaging in conversations with one of our Clubhouse members, I was able to figure out her interests and connect her to making activities. She loved video games, so I challenged her to create her own. She took an interest in physical computing and began creating projects on her own. She likes being in charge when working in groups, so I gave her the title of Project Manager. . . . Taking the time to learn about her interests allowed me to help her identify her strengths and build confidence so that she can proudly call herself a maker.

—Jaleesa Trapp, Tacoma Clubhouse, Tacoma, WA

CREATIVITY IS THE MAIN INGREDIENT

Making is taking any project and finding ways to expand on it. An example of this is taking something like a water bottle and making a beautiful flower out of it. A maker will also find ways to do more with that water bottle by finding a different object to make, like a plastic dress for a character made from scrap materials. I'm a huge fan of creating things out of recycled goods. Being a maker doesn't mean you have to go out and spend a bunch of money on high-tech electronics. Being a maker means taking what you have, like trash, and viewing it differently—think of it as material that you get to reinvent. A great example of this is plastic. Plastic is strong and very versatile, meaning it can have many uses. I challenge all makers to reinvent a piece of plastic material and see what you come up with!

—Audrey Maestas, Compadre Boys & Girls Club Clubhouse, Chandler, AZ

ADDITIONAL RESOURCES

- ▶ Tinkering Studio project ideas:
  tinkering.exploratorium.edu/projects
- ▶ Instructables DIY projects website:
  instructables.com
- ▶ Mentoring tips:
  youngmakers.org/mentoring-tips/
- ▶ Spin turntable to photograph creations:
  spin.media.mit.edu

Final project ideas...

Show and Share:

COMMUNITY SHOWCASE

In this final session, makers will share their projects with family, friends, and others in a community showcase event. They will have an opportunity to reflect on their process and celebrate their accomplishments.

SESSION GOALS

In this session, makers will

- Exhibit their work for others to see, try, and appreciate.
- Reflect on their projects, process, and progress.
- Imagine new possibilities for future projects based on the skills they have learned.

✔ GETTING READY

Plan for the community showcase well before the event. Engage your makers in planning the showcase, creating invitations, and decorating for the event.

Schedule this session so that your makers have some time to set up, make labels, and finish anything before presenting their projects. You'll also want to make sure there is

Community showcase (SCI-BONO CLUBHOUSE, JOHANNESBURG, SOUTH AFRICA)

Presenting a project at a Maker Faire (SAN RAFAEL CLUBHOUSE, SAN RAFAEL, CA)

dedicated time for everyone to document their work and share what they plan to present with each other before any guests arrive. Encourage makers to focus on their process of making, not just the end result. Invite friends and family members to come for the final hour of the session.

Poster for community showcase (FLAGSHIP CLUBHOUSE, MUSEUM OF SCIENCE, BOSTON, MA)

Preparing a display of projects (SACRAMENTO FOOD BANK & FAMILY SERVICES CLUBHOUSE, SACRAMENTO, CA)

Completing a project (FLAGSHIP CLUBHOUSE)

★ FACILITATION TIP

Looking Back, Looking Ahead

You can help your makers prepare for their Community Showcase by inviting them to reflect on their experiences through a series of questions about their idea, process, and what they learned.

Idea

▶ What did you make?

▶ What was the idea you started with?

▶ What made your idea change over time?

Process

▶ What was the hardest part?

▶ Did anything surprise you as you were working on this?

▶ Did you work with others on the project or help others? How?

Learning

▶ What is something you learned?

▶ How might you use what you learned?

▶ What is something you might like to try next?

Poster announcing showcase (BOYS & GIRLS CLUBS OF METRO WEST CLUB-HOUSE, FRAMINGHAM, MA)

Sharing a light-up project (THUNDERBIRDS BRANCH BOYS & GIRLS CLUB CLUBHOUSE, GUADALUPE, AZ)

RUNNING THE COMMUNITY SHOWCASE

Here are a few ideas to help you creatively showcase makers' experiences, projects, and processes.

▶ Create and play a slideshow of photos or videos of the earlier sessions.

▶ Offer hands-on "make and take" projects, such as paper circuit name badges, where makers lead demos.

▶ Create carnival-style booths where makers' friends and family can play with projects and are able to see the code behind the Scratch projects.

▶ Create a space for displaying Art Bot drawings and light painting photographs.

▶ Make the space inviting and interactive—for example, by creating a MaKey MaKey doorbell at the entrance.

▶ Make and wear homemade t-shirts that say "Ask Me About My Project!"

Community Showcase (FLAGSHIP CLUBHOUSE)

STORY FROM THE CLUBHOUSE NETWORK: A START MAKING! COMMUNITY SHOWCASE

The Clubhouse at the Boys & Girls Clubs of Metro West in Framingham, Massachusetts organized a final showcase for makers, friends, families, and other Clubhouse supporters. The makers prepared presentations of their final Open Make projects using poster boards to explain the materials they used, the challenges they faced, and what they hoped to make next.

For example, some of the young people shared musical projects made with the MaKey MaKey and Scratch. Here is a description from Clubhouse Coordinator Yarelis Garcia:

These two friends got together and worked on amazing hands-on MaKey MaKey projects incorporating copper tape, cardboard, brass fasteners, wires, construction paper, and magazine cutouts.

The purpose of their project was to re-create Guitar Hero using these materials and MaKey MaKey to create the physical instrument. They then worked with Scratch to create music. They also created an interface inspired by Dance Dance Revolution so that they can dance and play the guitar at the same time.

These girls have encountered many challenges and difficulties, but they have overcome many of them by taking their time researching and using creative methods.

Others at the Clubhouse have been inspired by their projects. Their Open Make project has impressed many of our guests and visitors. These girls now want to do making projects every day!

Building a dance project interface

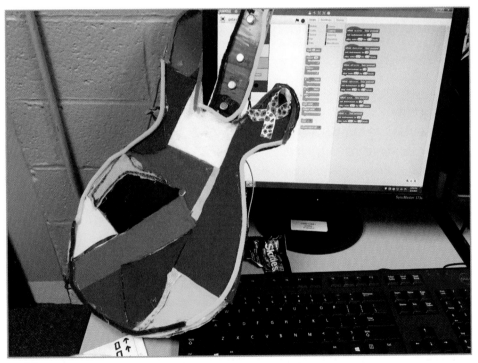

Building a guitar project

Project display boards

ADDITIONAL RESOURCES

▶ Start Making! video from the Clubhouse Casa de la Juventud: youtu.be/LKk5FOsbhs4

▶ *Creating in the Clubhouse: Tools for Conversations and Portfolio Development* developed by the Education Development Center's Center for Children and Technology: bit.ly/cct-clubhouse-toolkit

Poster for showcase (CASA DE LA JUVENTUD CLUBHOUSE, MORA, COSTA RICA)

Feedback I received when I shared my project...

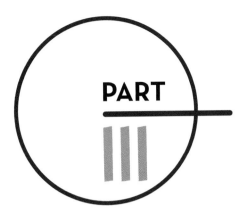

PART
III

Keep Making

Encouraging Youth
TO KEEP MAKING!

By facilitating Start Making! activities you can help young people develop a variety of creative competencies. In the process of making meaningful projects, young people learn new concepts and design skills. They become comfortable using a variety of tools and technologies to make projects. They gain confidence in their ability to bring their ideas to life and to persist through setbacks. In addition, they develop positive relationships through collaborating and helping others. Through this experience, they begin to see themselves as makers: as individuals who enjoy coming up with new ideas, creating personal projects, experimenting with new materials, and sharing their skills with others.

How can you inspire young people to continue making projects, trying new things, and going deeper into areas that they find interesting and meaningful? Here are some ways we've found to open up further opportunities for making and learning:

Youth encouraging others to Start Making! (PARROQUIA DE FÁTIMA CLUBHOUSE, PANAMA CITY, PANAMA)

Host an idea potluck. Organize a workshop where everyone brings an idea, shares the idea, and then works with others to make all the ideas come to life.

Find extraordinary uses for ordinary things. Encourage young makers to look for inspiration for further projects in things they see around them in their everyday environment.

Applying ideas from Start Making! to make new projects (YOUTH CONNECTIONS CLUBHOUSE, LISMORE, AUSTRALIA)

Engage families. Invite families to participate in creative exploration with young makers.

Celebrate by making. Celebrate holidays or special occasions (such as birthdays) by making projects in the spirit of the occasion.

Make new connections. Combine ideas and techniques from different Start Making! sessions to make new creations, such as soft circuits that make music.

Reverse roles. Dedicate a day in which young makers take ownership in planning and leading creative hands-on activities for others.

Go outside for a maker adventure. Pack up your Open Make box and take it outside for a maker field trip or picnic; create projects based on what you see around you.

Seek inspiration at events. Connect makers to broader maker events—such as Maker Faires, science festivals, and art performances and exhibitions—where they can find new inspiration and ideas.

Afterword:
A MESSAGE TO ALL MAKERS

—From Mitchel Resnick, professor and director of
the Lifelong Kindergarten group, MIT Media Lab,
and co-founder of the Clubhouse

Flagship Clubhouse, Museum of Science, Boston, MA

WELCOME TO THE MAKER COMMUNITY!

I hope you'll enjoy building projects using the ideas, tools, and materials described in this book. In the process, you'll learn many important things. For example, you'll learn math, science, and engineering concepts related to electrical circuits, conductivity, light diffusion, and computer programming. These concepts are important and valuable. But that's not the most important thing that you'll learn.

What's most important is that you'll learn how to think like a maker—and to think of yourself as a maker.

WHAT DOES IT MEAN TO THINK LIKE A MAKER?

It means that you know to start with a spark of an idea and turn it into a meaningful project.

It means that you know how to break down complex challenges into simpler parts.

It means that you know how to identify problems as they arise, to keep trying when things get difficult, and to come up with new strategies and approaches.

Collaborating on a project (SCI-BONO CLUB-HOUSE, JOHANNESBURG, SOUTH AFRICA)

It means that you know how to collaborate with others, to build on the work of others, and to share your ideas with others.

Learning to think like a maker is more important now than ever before. To thrive in today's rapidly changing world, you'll need to think creatively, reason systematically, and work collaboratively. And that's exactly what you learn as you work through the activities in this book.

But the activities in this book are just the beginning. The book is called *Start Making!* for a reason. Even when you've completed all the activities, you're still at the start of your journey as a maker. The challenge is to keep making!

WHAT ARE THE NEXT STEPS?

You might want to work on variations of the activities and projects in this book. What if you make one of your projects bigger? Or smaller? What if you combine two of your projects into one? Or what if you add some new materials to an existing project? How about modifying one of your projects to turn it into a gift for a friend?

Or you might try making a totally new type of project. For inspiration, take a look at other books, magazines, and websites featuring maker projects (like makezine.com). Or imagine a new project that connects with one of your favorite hobbies or interests.

It's often helpful to make and learn with others. There are a growing number of Makerspaces and Maker Faires where you can meet other makers. And there are online maker communities (like diy.org and scratch.mit.edu) where you can share your creations and see what others are making.

Learning to think like a maker is a lifelong process. You need to keep experimenting, taking risks, and trying new things. There are always new ideas to explore, new tools to use, new techniques to learn, and new projects to share. Let your curiosity be your guide.

The LEGO Group has a slogan that I love: "Joy of Building, Pride of Creation."

You'll make your best projects, and you'll learn the most in the process, when you embrace the joy of making, and when you share your projects (and your joy) with others.

Congratulations on joining the worldwide maker community! Keep making!

About the Clubhouse Community

The Clubhouse provides a creative and safe out-of-school learning environment where young people from underserved communities work with mentors to explore their own ideas, to develop skills, and to build confidence in themselves by using technology. The program was founded in 1993 as a collaboration between the Museum of Science, in Boston, Massachusetts, and the MIT Media Lab. To learn more, visit theclubhousenetwork.org.

Walk into a Clubhouse and you will find groups of young people producing and editing films, building robots, creating graphics and websites, making 3D models and animations, designing computer games, writing and recording music, and much more. In the process, they become excited about learning and about their own future.

Today, The Clubhouse Network is a global community comprised of 100 Clubhouses in 20 countries, which provide more than 25,000 youth per year with access to resources, skills, and experiences to help them succeed in their careers, contribute to their communities, and lead outstanding lives.

Based at Boston's Museum of Science, The Network supports community-based Clubhouses around the world by providing start-up support, professional development, new technology innovations, evaluation and assessment, access to an online community for youth, mentors, and staff, and more.

CORE VALUES

The success of the Clubhouse community depends on five core values:

Equal opportunity Empowering youth by granting full access to resources

Relationships Fostering healthy, respectful, and consistent relationships

Creative process Nurturing a community of lifelong learners and producers

Diversity Encouraging an inclusive environment that embraces the rich diversity of our communities

Hard fun Engaging in digital media, art, and STEM (science, technology, engineering, and math) tools to express, invent, and collaborate

For more than 20 years we have seen firsthand the impact of a safe, creative place that empowers young people to become more capable, creative, and confident learners. That is what makes us passionate about sharing the Clubhouse approach and the activities and projects in this book with others. We hope you too will see young people unleash their creative talents, build confidence in their own learning, and discover a unique voice of their own with which to express themselves creatively.

About the Authors

Danielle Martin Danielle served as knowledge manager and led the Start Making!
program across the global Clubhouse Network.
She started her Clubhouse career as a coordinator at the Boys & Girls Club in Charlestown,
MA, and previously served as an AmeriCorps
VISTA, supporting community-based media
and technology programs. After obtaining
a master's degree in City Planning from the
Massachusetts Institute of Technology (MIT),
she co-created a research group within MIT's
Center for Future Civic Media, the Department of Play, which mapped mobile technologies and youth activism methodologies. Now

**Danielle mentoring at a
Makers' Studio session** (BOYS
& GIRLS CLUB OF METRO WEST
CLUBHOUSE, FRAMINGHAM, MA)

she manages programs for Team4Tech.org, which is focused on advancing
21st century education in underserved communities by engaging U.S.-based
technology volunteers and IT solutions in collaboration with local nongovern-
mental organizations.

She started making by collaging on paper, then digitally in Photoshop, and by
hacking Polaroid prints, but now she enjoys mashing up photos on Instagram
or baking collages of treats for anyone who will eat them.

Alisha Panjwani Alisha is a designer and educator interested in exploring experiential and experimental ways of integrating story-telling, craft, wellness, play, and interactive technologies to create participatory learning practices. Her practice centers on nurturing children's creative confidence with new tech-nologies and encouraging their involvement in creative acts within their communities. She completed her master's degree in Media Arts and Sciences at MIT and worked as a research assistant in the Lifelong Kindergarten group at the MIT Media Lab. Before coming to the MIT Media Lab, she worked as a design and research associate at Project Vision, an international research initiative based in India that focuses on developing appropriate instructional strategies and technology-related tools that foster creative cognitive architectures in young children from urban poor communities.

Alisha mentoring during a Start Making! session (FLAGSHIP CLUBHOUSE, MUSEUM OF SCIENCE, BOSTON, MA)

She started making art projects even before her mother let her use a pair of scissors, but now she makes her own fashionable outfits for fun.

Danielle Martin and Alisha Panjwani

Natalie Rusk Natalie is a research scientist in the Lifelong Kindergarten group at the MIT Media Lab who develops programs that enable young people to build on their ideas and interests. She is one of the creators of the Scratch programming language and is co-founder of the Clubhouse program. She also researches the role of motivation and emotions in learning. She earned a master's degree at the Harvard Graduate School of Education and a PhD in child development from Tufts University. She enjoys making projects with others in workshops that combine recycled materials, LEGO bricks and motors, colored lights, and coding with Scratch.

Natalie introducing Scratch
(PHOTO BY JOI ITO)

Index